IMPLICIT LEARNING AND
TACIT KNOWLEDGE

OXFORD PSYCHOLOGY SERIES

1. *The neuropsychology of anxiety: an enquiry into the functions of the septohippocampal system*
   Jeffrey A. Gray
2. *Elements of episodic memory*
   Endel Tulving
3. *Conditioning and associative learning*
   N. J. Mackintosh
4. *Visual masking: an integrative approach*
   Bruno G. Breitmeyer
5. *The musical mind: the cognitive psychology of music*
   John Sloboda
6. *Elements of psychophysical theory*
   J.-C. Falmagne
7. *Animal intelligence*
   Edited by L. Weiskrantz
8. *Response times: their role in inferring elementary mental organization*
   R. Duncan Luce
9. *Mental representations: a dual coding approach*
   Allan Paivio
10. *Memory, imprinting, and the brain*
    Gabriel Horn
11. *Working memory*
    Alan Baddeley
12. *Blindsight: a case study and implications*
    L. Weiskrantz
13. *Profile analysis*
    D. M. Green
14. *Spatial vision*
    R. L. DeValois and K. K. DeValois
15. *The neural and behavioural organization of goal-directed movements*
    Marc Jeannerod
16. *Visual pattern analyzers*
    Norma V. Graham
17. *Cognitive foundations of musical pitch analysis*
    C. L. Krumhansl
18. *Perceptual and associative learning*
    G. Hall
19. *Implicit learning and tacit knowledge: an essay on the cognitive unconscious*
    Arthur S. Reber

Implicit Learning and Tacit Knowledge

An Essay on the Cognitive Unconscious

ARTHUR S. REBER
Professor of Psychology
Brooklyn College
City University of New York

OXFORD PSYCHOLOGY SERIES NO. 19

New York Oxford
OXFORD UNIVERSITY PRESS CLARENDON PRESS
1993

Oxford University Press

Oxford New York Toronto
Delhi Bombay Calcutta Madras Karachi
Kuala Lumpur Singapore Hong Kong Tokyo
Nairobi Dar es Salaam Cape Town
Melbourne Auckland Madrid

and associated companies in
Berlin Ibadan

Published by Oxford University Press, Inc.,
200 Madison Avenue, New York, New York 10016

Library of Congress Cataloging-in-Publication Data
Reber, Arthur S., 1940–
Implicit learning and tacit knowledge:
an essay on the cognitive unconscious /
Arthur S. Reber.
p. cm. — (Oxford psychology series; no. 19)
Includes bibliographical references and index.
ISBN 0-19-505942-5
1. Implicit learning. I. Title. II. Series.
[DNLM: 1. Cognition. 2. Unconscious (Psychology)
BF 315 R291i] BF319.5.I45R43 1993 153.1′5—dc20
DNLM/DLC for Library of Congress 92-49920

2 4 6 8 9 7 5 3 1

Printed in the United States of America
on acid-free paper

To JR, ER, and HA: the next wave.

Preface

This book is an attempt to present a coherent overview of the current state of affairs with respect to the study of implicit learning. It has, to be sure, an egocentric bias in that the focus tends to be on the work done in our lab; we've been looking at this mode of acquisition for longer than most, and it presents the story I know best. Chapter 1 introduces the key concepts that underlie the conception of implicit learning that I and my coworkers have used as the foundation of our examination of the cognitive unconscious. This chapter functions as a kind of "scene setter" in that several of the controversies that have emerged in the field are introduced and discussed and a short historical overview of the field is presented.

Chapter 2 is the data chapter. It presents a reasonably thorough review of the existing literature—with the recognition that with the rate that research is being carried out these days it is certain to be at least partly out of date already. The focus is on implicit learning and the tacit knowledge base that results from it. My primary aim here was to focus on the research that has enabled us to draw distinctions between the implicit, unconscious aspects of cognitive functioning and the explicit, conscious aspects. Although this is a topic that has been and surely will continue to be debated vigorously, my position is that implicit, nonreflective cognition has been shown to be dissociable from explicit, reflective cognition—and these differences lie at the heart of this chapter.

Chapter 3 is the theoretical chapter. Here a framework derived from basic principles of evolutionary biology is used to develop a model of implicit learning that provides an "in principle" argument that implicit and explicit modes of acquisition of knowledge must be dissociable. While this model is based on formal principles, it is presented as being somewhat "speculative" in nature. The reason for my (unaccustomed) caution here is that there has been relatively little experimental work directed at these issues for the simple reason that up to now no one has thought of taking this approach to the study of implicit learning. The point here is, while this evolutionary perspective turns out to provide an excellent account of much of the existing data as well as making several predictions about phenomena that have yet to be experimentally explored, little work has been carried out with the express purpose of testing this model. One of my hopes is that others will find this orientation sufficiently intriguing (or sufficiently annoying) that they will seek to examine its entailments carefully.

Finally, Chapter 4 deals with a rather extensive variety of topics that have emerged in the examination of the cognitive unconscious. Some of these are direct entailments of the work on implicit learning; others are topics of general importance in cognitive science independent of any particular focus. Some are primarily methodological, some theoretical, others philosophical. This chapter was supposed to be the "wrap it all up" chapter; alas, because each of the issues raised had so many extensions and entailments, it metamorphosed into a rather candid chapter. Some may be bothered by the occasional speculation that is engaged in; others excited by it. We shall see.

Finally let me wrap up this preface with a little personal note and a set of thanks to various folk. I sat down to write this preface and found myself feeling uncomfortable. Although this last piece of writing comes at the end when the hard work is done, I suddenly discovered that it carries a very real social burden. I now understand why everyone at all those award ceremonies babbles on and on, thanking everyone from their agents to their hairdressers. Until this moment I had been captured by my own arrogance ("*my* book, *my* ideas") and planning merely to jot down a two-line acknowledgment and be done with it. But that just won't do. There are, indeed, many individuals whose input, advice, support and, perhaps most important, criticism were essential.

I've been fascinated with the general problem of implicit learning for nearly thirty years now. Except for a few forays into projects in psycholinguistics and lexicography, it has been the issue that has commandeered my thinking. The earliest influences on my thinking came from Lee Brooks, who has been a good friend since we shared an office during our graduate school days back at Brown. I have had many people act surprised when they discover that Lee and I are friends since we seem to disagree so fundamentally on basic issues in cognitive science (one being that he does not like that label). I owe a special debt to him for calling the shots cleanly and honestly for nearly 30 years.

The book itself began with a simple suggestion from Donald Broadbent, with whom I have been corresponding for years and do hope to meet one of these days. In one of his long and typically insightful letters, he wondered whether I might consider putting together a monograph on implicit learning. He said he thought Oxford University Press might be interested. I was and they were, and I thank him for taking the initiative and for the many suggestions that spill forth from the pages of his correspondence.

The fraternity of scholars interested in exploring implicit cognitive processes has grown mightily in the past couple of years but it was not always so. There were a few like Pawel Lewicki and Bob Mathews who were intrigued by the implications of the phenomenon of implicit learning long before it became popular. I owe both of them more than they probably know.

The fraternity of scholars interested in criticizing implicit cognitive processes has, on the other hand, always been a large and active one from their early attacks on the New Look movement and subliminal perception to the sophisticated critiques of contemporary work in implicit learning and memory. Doing science is always a give-and-take affair, and any practitioner with any insight recognizes that in the final analysis they owe as much or more to their critics as to their supporters. Accordingly, I would like to thank both Don Dulany and Pierre Perruchet for keeping me honest.

On those occasions where I have decided that I had something to say that would be of interest to philosophers, I have found the sensible (and sobering) advice of Bill Bechtel and Bob McCauley of great help. We met when we were NEH Fellows at the interdisciplinary summer institute on Philosophy and Psychology of Mind in 1981 (or as we called it, "philosophy camp") and have remained close since then. One of the things that has pleased me most about the past decade or so in our field is that psychologists once again feel free to talk to philosophers—and both benefit.

There have been many of my colleagues at Brooklyn College who functioned as sounding boards for my ideas. I especially thank Matt Erdelyi, Neil Macmillan, and Don Scarborough for advice and criticism provided in equal measure.

In the last couple of years many students have worked with me on the problems outlined in this volume. I call our little home the Institute for Experimental Epistemology because I thought it a name both clever and impossible. Little did I know that the brilliant Warren McCulloch had coined the term many years earlier—well, it surely is good company to be in. Special thanks go out to the recent denizens of the IEE: Mike Abrams, Gary Cantor, Faye Fried, Toniann Genovese, Bernadette Guimberteau, Ruth Hernstadt, Andrew Hsaio, Mike Kushner, Selma Lewis, Max Locke, Lou Manza, and Bill Winter.

The folks at Oxford University Press were perfect, especially Joan Bossert. She hassled me (gently) at the right times and left me alone at the right times. What more can one ask of an editor.

Finally I want to thank my best friend, Rhianon Allen—who also happens to be my wife and occasional co-author. She says she didn't do anything to deserve thanks and doesn't want to be thanked. But it's my book and my preface and so she gets thanked.

Much of the research discussed in this monograph, not to mention the writing, was carried out under support from Grant BNS 89-07046 from the National Science Foundation and several grants from the City University of New York PSC-CUNY Research Award Program. Special thanks go to Joe Young and the people in what was the Human Cognition and Perception program of the NSF.

Brooklyn, N.Y. A. S. R.
June 11, 1992

Contents

IMPLICIT LEARNING AND TACIT KNOWLEDGE

1. Introductory remarks

The purpose of this chapter is to "set the scene," to establish a conceptual foundation for the rest of the book. I will introduce some of the key issues that have formed the conceptual focus for the work on implicit learning in which I, my students, and colleagues have been engaged over the past three decades. I will also try to lay out, as clearly as I can, the core problems under consideration in this book, characterize the notion of *implicit learning* as it has come to be used in the literature, and put this seemingly specialized topic into a general framework. I will also develop a loose theoretical model based on standard heuristics in evolutionary biology within which I have found it comfortable to think and, finally, lay the groundwork for the entailments and speculations concerning implicit cognitive processes and their general role in the larger scope of human performance. Finally, I will present a capsule history of interest in and work on the cognitive unconscious.

On learning

I want to make it clear from the outset that this will be, in large measure, a book about *learning*. A decade or three ago that would not have been particularly unusual; today it is a genuine rarity. Indeed, it is curious, given the pattern of psychological research over the course of this century that the topic of learning should be so poorly represented in contemporary psychology. A quick scan of some of the recent texts is revealing.

Most of the standard introductory textbooks cover learning in the context of conditioning and animal studies and run the standard historical gamut from Pavlov to Rescorla on one (classical) hand and Thorndike to Skinner and Herrnstein on the other (instrumental or operant) hand. Moreover, one comes away from a perusal of these works struck by the pattern of references in these chapters that is so different from those in the chapters on cognition and language. It is almost as though learning, as an area of independent study, is viewed as a historical topic. In the recent *Oxford Companion to the Mind* (Gregory, 1987), a general compendium of practically everything in contemporary (cognitive) psychology, the only acknowledgements of active, pure research on the topic since the 1950s are biofeedback and operant conditioning with animals.

The impact of what Baars (1986) has called the "cognitive revolution in psychology" on the study of the topics of learning and conditioning has been profound. One of the consequences of this dominance by the cognitive sciences has been that learning and conditioning are now typically interpreted within a cognitive framework. There has been, however, very little influence in the other direction—that is, not much of the work on conditioning has affected workers in the cognitive sciences. Herein lies a problem.

It seems obvious that this asymmetry has had unfortunate consequences. In Chapter 3, I will outline areas of conceptual overlap and show how, within an evolutionary framework, one can see learning and cognition as richly intertwining issues and not as two distinct fields with one dominating the other. The argument will be developed around the proposition that the basic principles of cognitive induction and abstraction on one hand and conditioning and associative learning on the other share a common process—the detection of covariation between events. The case for the generality of this process has been articulated by others (notably Holland, Holyoak, Nisbett, & Thagard, 1986, and Lewicki, 1986a). What I hope to bring to the analysis is the adaptationist's heuristic, the notion that such processes are best viewed within the context of evolutionary biology. The aim here, of course, will be to show that many aspects of modern cognitive science can profit from a recognition of the importance of much of the recent work on the topic of learning. The direction of influence here should properly go both ways.

Recognizing the importance of learning would have additional gains. Specifically, many of the important philosophical questions concerning the acquisition of knowledge would be returned to the forefront of the field. Generally speaking, when the cognitive revolution pushed the study of learning into the background, with it went much of the interest in the classic epistemic problems of knowledge acquisition with which philosophers and psychologists had struggled for decades. For the most part, contemporary cognitivists have been captured by the problems of knowledge representation and use rather than with an examination of its nature and acquisition.

Similar sentiments have been recently expressed by Glaser (1990) in his overview of the relationship between learning and instruction. Interestingly, he identifies the influence of the information-processing models as one of the primary reasons why learning has been neglected in recent decades. He specifically cites Newell and Simon's (1972) arguments that the study of performance, in an information-processing format, should take precedence over examinations of learning and development as a critical element in the drift away from learning. This analysis fits with the one I have suggested in that the Simon and Newell research program also gives formalizations a high priority and invites the building of theories of knowledge representation rather than knowledge acquisition. Glaser is rather sanguine about the reemergence of learning theory and argues that increased interest in the theory of instruction is, in large measure, responsible. One can only hope his analysis of the situation is correct.

An additional hopeful sign within cognitive psychology and the cognitive sciences is the resurgence of interest in the processes of knowledge acquisition in the guise of the development of artificial-intelligence (AI) systems. Of particular interest are those that are capable of carrying out induction and discovery routines, such as the parallel distributed processing (PDP) models of McClelland and Rumelhart (McClelland & Rumelhart, 1986; Rumelhart & McClelland,

1986), Holland et al.'s (1986) rule-based production system for induction, and John Anderson's ACT* model, which has been recently applied to problems such as knowledge acquisition and compilation (Anderson, 1987) and knowledge transfer (Singley & Anderson, 1989). Anderson's program is, of course, significant because of all of the AI-oriented models his is the one that has been designed to attempt to handle the varied issues of instruction, the acquisition of skills, and the development of expertise. Although the model I will develop in this book does not have anything like the formal instantiations of these systems, the conceptual similarities are strong, and I find this encouraging.

However, my focus will be on a particular kind of learning, specifically what was dubbed back in 1965 as *implicit learning* (Reber, 1965). Implicit learning is the acquisition of knowledge that takes place largely independently of conscious attempts to learn and largely in the absence of explicit knowledge about what was acquired. One of the core assumptions of our work has been that implicit learning is a fundamental, "root" process, one that lies at the very heart of the adaptive behavioral repertoire of every complex organism. As such, it is the concept of implicit learning that can, in principle, encompass the early work on learning and conditioning and tuck it into a more general epistemic framework along with the more contemporary analyses of learning, induction, and discovery. Chapters 2 and 3 will cover much of this argument.

On nativism and empiricism

The same standard sources that reveal the shifting pattern of emphasis on learning and cognition also reflect an accompanying tilt away from the classical empiricist's point of view and toward the nativist's. Along with the deemphasizing of learning, there has been a clear resurgence in nativistic theorizing exemplified most notably by the work of people such as Chomsky (1980, 1986) and Fodor (1983). There are interesting and controversial historical patterns in these shifts. It seems reasonably clear that these two paradigms are like the opposite ends of a seesaw. The flowering of empiricism typically accompanies advances in learning theory and results in a lessening of the influence of nativist thinking. Neglect of the processes of knowledge acquisition correlates with resurgences in nativistic theory and increases in the attractiveness of the presumptive a priori.

There are many ways to view the long (and complex) history of the shifting fortunes of these points of view. One standard analysis argues that the failures of empiricism simply show the truth of nativism. As this argument goes, the empiricists' inability to produce a theoretical framework that could explicate the acquisition of complex knowledge (such as that represented by a natural language) should be viewed as support for the alternative notion that such knowledge is not "acquired" at all but given a priori. Here, so the line of argument continues, the deemphasis on learning comes not from any subtle shift in priorities within the field but because of a kind of "rats abandoning the sinking ship" effect. If empiricism is bankrupt, then all sensible scholars should cleave to nativism.

Another line of analysis, somewhat kinder to the empiricists, maintains that the nativists are simply too impatient. They view plateaus in the progress of the study of the processes of knowledge acquisition as evidence of ultimate failure rather than the waxing and waning that characterizes normal science. The standard line here, to which I for one subscribe, is that a return to serious work on the problem of acquisition is just about due and that it will provide a number of intriguing insights into the generality of the processes of learning, induction, and utilization of complex knowledge.

In the end, however, the debate hinges on one critical issue. Empiricism cannot be defended without a theory of learning. When empiricism with its attendant environmentalism emerged in a mature form during the seventeenth century, the coherence of its challenge to Cartesian rationalism was dependent on fulfilling certain obligations concerning the processes of acquisition. Rationalism, possessed of a convenient nativism, simply assumed that the necessary epistemic "machinery" for making sense of the world was given, a priori. In dismissing such Cartesian assumptions, empiricists were obligated to develop mechanisms to explain how all this knowledge "got there." Associationism, with its rich Aristotelian foundations, of course, became the dominant theory. It still is—although the modern versions of associationism bear little semblance to the original. It should come as no surprise that many of the recent defenses of nativism focus on the supposed fragility of associationism (Fodor, 1983).

It is no mere coincidence that with the lessening of interest in the study of learning there was a return to Cartesian rationalism and nativism. The British Empiricists' obligation still holds true: if you don't provide a sensible characterization of how knowledge is acquired, it is easy to simply assume that it was "born in," which is exactly what Fodor and Chomsky have maintained. In Chapter 4 these points will be developed in more detail. A basic presumption will be that there are good reasons for resisting nativism; one of the best antidotes to nativism is a systematic and vigorous focus on the processes of acquisition. A second proposition will be one introduced some time ago (Reber, 1973) to the effect that there are variations in contemporary nativist theorizing and that each entails a different kind of model of mind. Not all forms of nativism inhibit the development of or interest in acquisition processes.

I should make it clear at the outset that this issue of nativism versus empiricism is a much larger one than can (or should) be tackled in a book such as this. It is one of our classical philosophical conundrums. It has not succumbed to any particular form of argument over the past couple of centuries, and I don't expect it to do so now. However, it is an issue of no little importance and, I will argue, relevant for the central thesis of this book. Moreover, it is an issue that has implications not just for how the cognitive sciences tilt but also for political and sociocultural values. I will be presenting arguments in support of a strong environmentalism that developed around the thesis first put forward by Dewey—that the empiricist position is the proper *default* position and that nativism should only be adopted when the evidence against empiricism becomes overwhelming.

On evolution

Evolutionary theory has proven over the decades to be one of the most effective "engines of discovery" in modern science. In psychology, it has proven itself to be a theory with truly remarkable heuristic value, focusing research on topics of central importance such as function, adaptation, individual differences, and phylogenetic analyses and promoting integration between psychology and related sciences. Much of the thinking underlying this book's characterization of implicit learning and tacit knowledge has been focused by considerations of evolution. As argued in Chapter 3, these considerations virtually force the model of implicit learning proposed. The critical observations are reasonably straightforward.

First, consciousness and phenomenological awareness are recent arrivals on the phylogenetic scene. Hence, consciousness and conscious control over action must have been "built upon," as it were, deeper and more primitive processes and structures that functioned, independently of awareness. On these grounds it is assumed that the processes studied under the rubric *implicit learning,* operating independently of consciousness, are more primitive and basic that those that are dependent, in some measure, on consciousness and conscious control. The argument, already hinted at, is that there are remarkably close ties between the typical experiment on implicit learning and the standard study of conditioning. The commonality lies in the detection of covariation between events, which, I will argue, is the deep principle in processes as seemingly disparate as classical conditioning and implicit learning. Moreover, this conceptual parallel can be shown to hold, even though on the surface the implicit learning experiment appears to be one of abstract induction and the conditioning experiment one of simple association.

Second, one of the standard heuristics in evolutionary theory is that phylogenetically older and more primitive structures will display telltale classes of properties different from the more recently evolved. One of these is that the structures with greater antiquity tend to be more robust and resilient, less prone to disruption of function than the newer. Therefore, we would expect to see implicit cognitive processes show greater resistance to interference from neurological insult and clinical disorder than the explicit processes. There is a large and rapidly growing literature on implicit learning and memory that supports this analysis; it will be reviewed in Chapter 3.

Third, the evolutionarily more ancient implicit functions of the cognitive unconscious should show a tighter distribution in the population than the more recently emerging explicit and the conscious—we should expect to find fewer individual differences between people when implicit processes are in use than when explicit processes are. This distribution proposition has two avenues of reasoning behind it. First, part of the standard model of evolution holds that the more primitive phylogenetic structures and functions, being the successful outcome of eons of adaptation, display less variation from individual to individual. Hence, those structures and functions responsible for implicit process will be displayed by a basic functional level of performance that is as close to being universal as any known cognitive process. Second, the educational programs and theories of instruction that dominate pedagogic practice in our society concentrate almost entirely on the explicit and overt functions. Such a focus in a society with the inequalities of ours should lead to an increase in the population variance on virtually any explicit cognitive function one chooses to measure. Both lines of argument here are, putting it mildly, controversial; the issues will be developed more fully in Chapters 3 and 4.

Fourth, there should be a reasonably clear relationship between the point along the phylogenetic tree where a particular property or function evolved and the degree to which we are conscious of its form and content. That is, we would expect to find that the more primitive a function is shown to be, the more refractory to consciousness it will be. There has not been as much work on this proposition as on some of the others, but what there is will be introduced below.

For these, and some other reasons introduced in Chapter 3, I see the study of implicit learning as a kind of evolutionary exercise. It is surprising that evolutionary theory as a heuristic device has been so rarely exploited in psychology in recent years. It is indeed an engine of discovery, and I, for one, feel comfortable working within its confines.

On methods and measuring of the contents of consciousness and the unconscious

There are many methodological booby traps awaiting anyone who ventures to study unconscious processes. They have been well and often vigorously debated in the literature, and they will be reviewed in Chapter 2 and discussed further in Chapter 4. However, this topic needs to be approached with caution; I have a deep suspicion that the problems are even more complex than most have realized.

The difficulty is that determining just what is conscious and unconscious at any point is not merely a theoretical or an epistemological problem. It is, first and foremost, a measurement problem. Some of the mental content that a typical subject acquires in the typical experiment on implicit learning may be very much a part of an individual's awareness, but determining this is going to depend on the developing a method for assessing conscious content. And, it should come as no surprise, different techniques will yield different measures. The point to be developed here is that questions of the epistemic status of implicit and unconscious knowledge will be dependent on a simple inequality (Erdelyi, 1986)

$$\alpha > \beta$$

where α is information available to the unconscious processing system and β is that available to the conscious system. Since all of the entailments of the evolutionary considerations give the implicit systems priority, the key to uncovering the cognitive unconscious will be found in measures of mental content held consciously (β) that yield lower estimates when compared to those made of the mental content held outside the purview of consciousness (α). Just how such measures are made and what kinds of psychometric defenses of them are provided will prove to be extremely difficult yet critical areas of investigation.

The "solution" to the problem, which will be proposed and defended in Chapters 2 and 3, is that the unconscious should become the default condition. That is, rather than putting the burden of demonstration on those who claim that a particular process was unconscious or that a given knowledge base was tacit, the burden of proof should go to those who argue that it was, in fact, conscious. In short, I will assume "the primacy of the implicit." We would do well to recognize that it is actually more surprising that any function is conscious than unconscious. Like many of the above proposals, this one is going to take a good bit of argument to be convincing.

On intelligence and instruction

Although the exploration of the cognitive unconscious was initiated from the point of view of a "pure" researcher, over the years I have begun to suspect that understanding these unconscious, covert systems could have legitimate and maybe even important applied aspects, particularly in the area of instruction. In various places I will discuss some real-world skills for which there are good reasons for believing are dependent on implicit functions. I will also bring to bear a growing literature indicating that the relationships between instruction, performance, and intelligence will be better understood if the distinctions and interactions between implicit and explicit learning are taken into consideration.

This will involve an overview of the historical issues in the very definition of intelligence—an outline of the hypothesized connections between implicit learning, implicit memory, tacit knowledge, and intelligence. The thrust of the proposal here will be that intelligent behaviors, long associated with the overt and conscious domain of cognitive functioning, are better seen as the result of both implicit and explicit capacities.

A note on terminology

Back in the 1960s when we began working on the general problem of implicit learning, I felt a distinct reluctance to use the term *unconscious* to characterize the phenomena appearing with regularity in the laboratory despite the obvious fact that the processes we were examining were indeed just that. The reasons were less than mysterious. There was (and to some extent, still is) considerable semantic "spread" from the psychoanalytic community's use of the term; for a young researcher in the as yet unacknowledged cognitive revolution, any suggestion of conceptual familiarity here was seen as seriously compromising to one's stature. Moreover, experimental psychology had just gone through an embarrassing episode with the related problems of subliminal perception and perceptual vigilance and perceptual defense, and psychological distance seemed wise. Other related terms like *incidental learning* and *learning without awareness* had already been co-opted and, anyway, were really different kinds of phenomena. The neutral term *implicit learning* was chosen simply to differentiate the processes from the *explicit learning* research on concept formation and categorization that others, such as Bruner, Goodnow, and Austin (1956), were doing.

It is well worth noting that recent work has returned many of these richly controversial topics to positions of respectability within the community of cognitive scientists (see, e.g., Erdelyi, 1985). Some of the relationships between the cognitivist's approach to implicit learning and tacit knowledge and the recently developed cognitive interpretations of psychoanalysis are most tantalizing (see any number of contributions to Bowers & Meichenbaum, 1984).

A rapid historical overview

The recent interest in the issue of implicit learning has been accompanied by a renewed curiosity concerning the actions and functions of the cognitive unconscious, considered broadly. This section provides a historical overview of the avenues of research that appear to me to have been significant in bringing about this development. I do not mean this to be a thorough history of the cognitive unconscious. In many ways it is a bit too soon, particularly since in its modern form it is less than three decades old (see Schacter, 1982, 1987, for historical overviews of much of the earlier work as it pertains to the issue of implicit memory). However, I do think that we have enough distance to see several, reasonably obvious elements that have, over the past thirty-odd years, contributed significantly to the current understanding of the cognitive unconscious. Taking them in turn:

The discovery of implicit learning

The earliest studies touching directly on the acquisition of complex information without awareness were those we carried out in the middle and late 1960s using the artificial-grammar learning procedure and a variation on the probability-learning experiment (Reber, 1967a, 1967b, 1969; Reber & Millward, 1965, 1968). Since these form the empirical core of this book, they are reviewed in considerable detail in Chapter 2.

There had, of course, been earlier research programs that toyed with various aspects of unconscious cognitive functions. However, for the most part, these approaches did not directly address issues of current interest. Notably absent in most of this early work is a concern with the question of the relative contributions of unconscious and conscious control over the acquisition process. For example, Clark Hull's (1920) early work on the learning of the structure of Chinese-like ideographs identified the process of concept formation by abstraction of common elements, a process still regarded as important (Medin, 1989). Hull's characterization of concept acquisition has some interesting similarities with perceptual learning, a process that takes place largely in the absence of awareness of the rules governing perceptual displays (E. J. Gibson, 1969; J. J. Gibson, 1979). Hull's work, while it implicated variables we now regard to be of considerable importance, had little impact on the field at the time; the Gibsons' research program became extremely influential in the study of perception but had little impact on learning. Hull, of course, abandoned these "cognitive" issues and shifted his focus to the study of motivation and reinforcement—and thereby embarked on a research program that made him one of the most influential behaviorist theorists of the 1940s and 1950s.

There was also a substantial data base built up during the middle decades of the century on the twin issues of *learning without awareness* and *incidental learning* (see here, Greenspoon, 1955; Jenkins, 1933; Thorndike & Rock, 1934). There is a touch of irony here in that much of this clearly cognitive research was initiated by those with a behaviorist orientation. It was carried out as part of the examination of the problems of motivation and reinforcement in learning—the very issues that Hull, having left behind matters of mind, had made the central concerns of a behaviorally oriented experimental psychology. Most of this research involved the use of the repeated trial design, in which subjects were presented with complex stimuli (usually linguistic) and were differentially reinforced for making particular responses. The substantive issues surrounded the question whether subjects could show behavioral evidence of having learned something about the associations between stimuli and responses to which they had been exposed without being aware of the S-R links.

Unfortunately, for the most part, these researchers struggled with problems of methodology and disputes over the degree to which incidental learning was truly incidental and just how much awareness there was in experiments on learning without awareness (for overviews here, see Eriksen, 1960, and Osgood, 1953, Chapter 10). The questions about counterbalancing roles of consciousness and unconscious mechanisms in complex learning were, of course, rarely explicitly raised. When they were it was often to point out the methodological inadequacies that seriously compromised this work (see Brewer, 1974).[1]

The early implicit learning experiments that focused specifically on the issue of unconscious acquisition of complex knowledge were run using complex, rule-governed stimuli generated by a synthetic, semantic-free, Markovian grammar. In the typical study, subjects memorized strings of letters in the synthetic language and were later tested for their knowledge of the rules of its grammar by being asked to make decisions concerning the well-formedness of novel strings of letters (see Reber, 1967a, 1989a). Unlike the materials in the earlier work on learning without awareness and incidental learning, which consisted typically of word lists and word associations, the stimuli here were composed of unpronounceable sequences of letters whose order was determined by arbitrary rules.

[1] There is an intriguing paradox here. As the behaviorist influence faded so did the interest in these essentially cognitive issues. Yet, the recent concerns with the cognitive unconscious actually reflect in subtle ways many of the points first introduced by behaviorists, particularly the notion that learning occurs independent of the learner's awareness of the process.

The use of arbitrary, semantic-free stimulus domains ensured that their underlying structures would not be known by the subjects prior to entering the laboratory. Optimally, implicit learning should be examined in a setting in which the acquisition process is unlikely to have been contaminated by previous learning or preexisting knowledge. As such, these experiments were specifically designed to explore the phenomenon of implicit learning, the process by which people acquired complex knowledge about the world, independent of conscious attempts to do so. They were also presented in a manner that clearly demarcated them from a number of other research programs of that era, particularly those of Bruner (Bruner et al., 1956) and Miller (Miller, 1967; Miller & Stein, 1963) that were based on the examination of overt and conscious hypothesis testing.

This early work was also motivated by a desire to examine empirically some of the classic philosophical questions of epistemology, including the acquisition and representation of complex knowledge. Among those thinkers whose interest in epistemological issues influenced these early grammar learning studies were two social philosophers: Michael Polanyi (whose original training, interestingly, was in medicine and physical chemistry), who had argued effectively for the importance of *tacit* knowledge, knowledge whose origins and essential epistemic contents were simply not part of one's ordinary consciousness (Polanyi, 1958), and Friedrich von Hayek (1962), the conservative economist who had put forward some elegant and controversial speculations concerning the necessity for deep rules and other rich mental representations to be held in a kind of "supraconscious" that was not available for ordinary conscious inspection.

As an aside here, it strikes me as curious that during this era other social scientists felt at liberty to explore these aspects of mind while psychologists, even those whose announced interests were in the study of cognitive and mental phenomena, seemed to be distinctly uncomfortable with such notions. I suspect that the problem stemmed, in no small measure, from our unfortunate reluctance to interact with philosophers, a prejudice only recently overcome. While we traditionally paid homage to our academic progenitors, few psychologists received much in the way of philosophical training past our undergraduate years. The resulting closed-mindedness also contributed to the rather narrow perspective that many experimental cognitive psychologists had until the gradual emergence of the truly interdisciplinary cognitive sciences in the 1980s. I will have more to say on this tendency toward encapsulation later. We have already seen it emerging in the often quaint terms introduced when the concept of the cognitive unconscious became an issue, and it will become even more obvious in Chapter 3 where the current and serious lack of interaction between workers in some of the subareas of implicit cognition has proven to be a problem.

As the research evolved, implicit learning came to be viewed as a rather general information acquisition process. By the middle 1970s it was being characterized as a situation-neutral induction process whereby complex information about any stimulus environment may be acquired largely independently of the subjects' awareness of either the process of acquisition or the knowledge base ultimately acquired. Most of this work was still being carried using the artificial-grammar (AG) learning procedure, although the paradigm was now more widely used (Brooks, 1978; Gordon & Holyoak, 1983; Howard & Ballas, 1982; Morgan & Newport, 1981; McAndrews & Moscovitch, 1985; Reber & Allen, 1978; Reber, Kassin, Lewis, & Cantor, 1980; Reber & Lewis, 1977). In addition, a number of other studies were run that employed a variation on the classic probability-learning (PL) procedure in which subjects had to predict which of several events would occur when the events followed any of a variety of probabilistic sequences (Millward & Reber, 1968, 1972; Reber & Millward, 1965, 1968, 1971). Details on these experiments will be provided in Chapter 2.

The examination of implicit learning continued using these two laboratory settings through the next several decades, although without stimulating much interest among cognitive psychologists outside of those who were concerned with the specific problem itself. The emergence of the interest in the larger questions of the cognitive unconscious would likely not have developed were it not for the following other factors.

The rediscovery of the nonrational

One of the unspoken (implicit?) elements of the period during which the early implicit learning work was being carried out was that humans are rational and logical and they reach conclusions and make decisions based on coherent patterns of reflection and analysis. At least this was the general point of view within the early decades of the nascent field of cognitive psychology (see Baars, 1986, for a cogent history of this era). During the 1970s, however, it became increasingly apparent that people do not typically solve problems, make decisions, or reach conclusions using the kinds of standard, conscious, and rational processes that they were more-or-less assumed to be using.

People appear to be, generally speaking, *arational*. It is not so much that we act in ways that do violence to rationality—although history shows no shortage of examples of such. The important insight was that, when people were observed making choices and solving problems of interesting complexity, the rational and the logical elements were often missing. It was not so much that decisions were being made that were irrational, it was rather that decisions were being made on the basis of processes that simply failed to take into consideration rational elements. Moreover, importantly, people often did not seem to know what they knew nor what information it was that they had based their problem solving or decision making on. This theme developed from a number of interdependent approaches to the study of human judgment, most significantly the work of Kahnemann and Tversky, of Richard Nisbett and his co-workers, and of Ellen Langer and her colleagues.

Kahneman and Tversky, in a now-classic series of studies, showed that issues of rationality and logic were largely independent of decision making and were often "replaced" by less than optimal heuristics. Often, these nonoptimal cognitive operations were displayed in the very contexts where one would imagine them to be most compellingly employed—for example, the statistician making a decision that violated Bayesian principles or a physician making inappropriate choices in triage-type settings (see the various contributions in Kahneman, Slovic, & Tversky, 1982, for an overview of this work).

An additional avenue of research emerged from social psychologists, who were examining how people made and justified "real-world" decisions. The work of Nisbett and his colleagues at the University of Michigan (Nisbett & Ross, 1980; Nisbett & Wilson, 1977) was particularly important. They directly addressed the notion that there were important cognitive lacunae between the (explicit) knowledge that we thought we used to make decisions and control choices, the (implicit) knowledge we actually used, and our differential capacities to articulate these kinds of knowledge.

In a related series of studies, Ellen Langer and her colleagues at Harvard showed that people frequently functioned in ways that were, to use their term, *mindless* (Langer, 1978; Langer, Blank, & Chanowitz, 1978). In situations where people appeared to be acting according to explicit and consciously developed inferences they were, in fact, drawing on implicit knowledge systems about which they had little or no awareness. Under such circumstances, people provided justifications for their behavior that were clearly at variance with what they had actually done. A substantial literature has grown in this allied area of social cognition over the past two decades. See Wegner and Vallacher (1977) for an overview of the earlier work, and the various contributions in Uleman and Bargh (1989) for the more recent advances.

This period represented an interesting episode in the rediscovery of the unconscious. There had been, of course, the psychoanalysts' earlier assault on rationality, yet with few exceptions, cognitive psychologists remained relatively uninfluenced by that message (see Erdelyi, 1985, for a conspicuous exception). With the development of this avenue of research from areas more closely aligned with the experimentalists' tradition, the notions of implicit learning and tacit knowledge were finally making their way into cognitive psychological thought, although often in any of a number of disguises. It was now abundantly clear that a good deal of cognitive "work" went on independent of normative notions of the rational and the logical and outside of the range of awareness.

The generality of implicit learning

Perhaps stimulated by this growing interest in implicit mentation, a number of researchers from diverse backgrounds and often with rather different agendas began to examine the phenomenon of implicit learning in an increasingly widening range of empirical settings.

Lewicki and his colleagues at the University of Warsaw and later at University of Tulsa, working initially from the point of view of social psychology and personality theory, reported results in a series of experiments that strongly paralleled those from the early synthetic language and probability learning studies. They found implicit acquisition of often extremely complex forms of information in experiments ranging from those on the perception of rule-governed spatial locations of stimuli (Lewicki, Czyzewska, & Hoffman, 1987; Lewicki, Hill, &

Bizot, 1988; see also here, Nissen & Bullemer, 1987; Stadler, 1989) to the processing of social information and personality characteristics (Lewicki, 1986a), and the development of self-perpetuating biases for coding information about social situations and personality characteristics of target persons (Lewicki, Hill, & Sasaki, 1989).

At Oxford University, Berry, Broadbent, and their colleagues discovered similar patterns of acquisition of covert knowledge in an extended series of experiments that explored how individuals developed the capacity to control complex environments, such as a simulated production plant or a socially interactive "computer person" (Berry & Broadbent, 1984, 1987, 1988; Broadbent & Aston, 1978; Broadbent, FitzGerald, & Broadbent, 1986). Similar findings by Mathews and his co-workers (Mathews, Buss, Chinn, & Stanley, 1988; Mathews, Buss, Stanley, Blanchard-Fields, Cho, & Druhan, 1989; Stanley, Mathews, Buss, & Kotler-Cope, 1989) extended and refined these conclusions.

In Chapter 2 details of these studies and a deeper interpretation of their implications will be discussed. All in all, these reports established the position that the implicit acquisition of complex knowledge was not an isolated phenomenon, but one of considerable generality and, as some were beginning to suspect (see Lewicki, 1986a; Lewicki & Hill, 1989; Reber, 1989a, 1989b, 1992a, 1992b), of genuine importance in human cognition.

Automaticity and procedural knowledge

By the middle of the nineteenth century, it was well understood that many complex perceptual processes were dependent on operations that lay outside of consciousness. The most effective proponent of this position was Helmholtz, as represented in his doctrine that perception was dependent on the process of "unconscious inference" (Helmholtz, 1867). Helmholtz's arguments, along with similar points of view espoused by Carpenter (1874), Hering (1920), and Ebbinghaus (1885), were sufficiently strong that even such recalcitrants as James, who had referred disparagingly to such proposals as "mind-stuff theory," were ready to admit that automatic and unconscious encoding existed (James, 1890, chapter 11).

After a hiatus of some decades, the relevance of these varieties of nonconscious encoding processes became apparent with the development of interest in the related problems of automaticity and procedural knowledge. The interest in automaticity grew out of the work of Hasher and Zacks and their colleagues (see Hasher & Zacks, 1984, for a review) that showed that such fundamental operations as encoding the frequency and location of objects and events in the environment took place automatically and largely without awareness of the encoding process. Hasher and Zacks also argued that this encoding process was a primitive and fundamental cognitive process and, as such, was relatively unaffected by variables such as age, developmental level, IQ, and affective state, which normally have considerable impact on cognitive processing. Although there has been some dispute over just how robust the specific process of frequency encoding is, the general proposition that automatic processes are different in fundamental ways from the consciously controlled, effortful processes is generally accepted (see Kahneman & Triesman, 1984).

Automatic processes are classic examples of the actions of implicit systems; they lie outside of consciousness and conscious control, they are "engaged" by events in the environment and not by intentions, and they are highly efficient in that they require few attentional resources. Importantly, they remain largely independent of factors that have significant impact on reflective and conscious encoding processes. These properties will turn out to be important for the evolutionary model developed in Chapter 3.

The other line of highly influential research emerged from the work of John R. Anderson and his colleagues on procedural knowledge (see Anderson, 1976, for the early approach, and 1983, for a more developed theory). Anderson's key distinction is that between declarative knowledge, which is knowledge that we are aware of and can articulate, and procedural knowledge, which is knowledge that guides action and decision making but typically lies outside of the scope of consciousness. Anderson's view is that virtually all interesting complex human skills are acquired in a characteristic fashion. They begin with the labored, conscious, and overtly controlled (declarative) processes of the novice that gradually give way to the smooth, unconscious, and covertly controlled (procedural) processes of the expert. This ordering of processes appears, at first, to be at variance with that presented by the standard theory of implicit learning in which the initial phases of acquisition are marked by a lack of consciousness. The two approaches actually can be shown to be complementary in that the domains of applicability of the two models are different. Implicit learning theory says little or nothing about skill learning. These issues will be discussed in more detail in Chapter 4.

These various avenues of research contributed in significant ways to the general interest in nonconscious cognition. Importantly, like the early research on implicit learning, the work on automaticity and declarative and procedural knowledge also provided empirical support for some of the classical philosophical problems in epistemology, again strengthening our ties with philosophy. The concept of automaticity is closely aligned with Polanyi's (1958) notion of tacit knowledge, and the distinction between declarative and procedural knowledge is neatly analogous with Ryle's (1949) distinction between "knowing that" and "knowing how."

This bringing of an empirical data base to bear on problems of epistemology was an important element in the development of a truly interdisciplinary cognitive science during the past decade. As I argued earlier, it forced cognitive psychologists to pay attention to issues of mind that philosophers had long regarded as critical to sensible theories of cognition. Moreover, it encouraged philosophers to begin to monitor more closely the work in cognitive psychology. As Bechtel (1988) has pointed out, philosophical inquiry has traditionally been suspicious of experimentalism, preferring rich and detailed argument about plausibility to "hard" data, which of necessity must be collected under resource-limited, controlled conditions. Yet this blending of experimentalists and epistemologists has proved exciting and productive and broadened cognitive psychology's domain of investigation.

Implicit memory

During the past decade, an influential, parallel line of research developed, one concerned with the examination of implicit *memory.* This approach differed from that taken to implicit learning in that questions about the acquisition of knowledge were seldom raised; the focus was on the processes of storage and retrieval of knowledge. Generally speaking, implicit memory is taken to have been displayed whenever a subject evidences, by some indirect or implicit measure of performance, that there was a memorial residue of an earlier experience in the absence of any comparable phenomenological sense of the previous experience.

Interest in unconscious memory has, perhaps not surprisingly, a long history (see Schacter, 1987). It ranges from various early philosophical treatments of memory such as Descartes' pre-Freudian speculation (cited by Perry & Laurence, 1984) that unpleasant early life experiences affected one's adult life, even though there were no conscious memories of the episodes, to numerous explorations of psychodynamically oriented scholars such as Janet, Bergson, and Freud (see Ellenberger, 1970, and Erdelyi, 1985). Serious empirical study of implicit memory, however, is relatively recent and, for reasons that will become apparent, a substantial amount of the work has been carried out on patient populations with various psychiatric and neurological memory impairments. During the past two decades, the variety of studies and the range of populations examined in an attempt to gain some understanding of implicit memory has been growing seemingly without bound. In what follows, the work with normal subjects will be reviewed briefly first, then the work with the variety of special populations will be presented within the larger topic of the robustness of implicit cognitive systems.

Among the first experiments that proved to be relevant here were those involving subliminally presented information. In these subliminal perception studies the stimuli were presented using any of a number of techniques designed to ensure that the material was not consciously encoded, including tachistoscopic presentation, masking, degrading, shadowing, and parafoveal presentation. For example, in what has become a classic experimental procedure, subjects are shown rapidly presented, masked visual displays designed to make it unlikely that they will be able to determine their identity. Under such conditions subjects have been shown to exhibit the memorial residue of presented material by their choices in semantic or preference tasks even though they are not aware of having been presented with the material and cannot select the "old" stimuli on two-alternative, forced-choice recognition tasks (Kunst-Wilson & Zajonc, 1980; Marcel, 1983a; Seamon, Brody, & Kauff, 1983; Seamon, Marsh, & Brody, 1984).

Other studies, some of which had originally been designed to explore the manner in which lexical information was stored in memory, produced findings that also implicated an implicit memorial system. Many of these experiments involved the use of the *repetition-priming effect* in which there is increased facilitation in the processing of stimulus material presented previously, independent of explicit memory for that material. This general finding has been reported in a large number of settings, including the lexical-decision task (Scarborough, Gerard, & Cortese, 1979), perceptual identification (Jacoby & Dallas, 1981), and word-stem completion tasks (Graf, Mandler, & Haden, 1982; Tulving, Schacter, & Stark, 1982). The interest in priming and implicit memory has become rather intense in recent years because the evidence for implicit memory using these procedures is so strong (see any number of the contributions in Lewandowsky, Dunn, & Kirsner, 1989).

In the past two decades, a large literature has grown up around the topic of implicit memory. Much of the recent work has been reviewed by Schacter (1987) and Tulving and Schacter (1990). All in all, the findings of the many studies on implicit memory parallel those on implicit learning. These commonalities will be discussed later in Chapter 3 in the context of the evolutionary model; there, it will become clear how strong the parallels between the two are. In Chapter 4 I will present a plea to bridge the gulf between these two areas of investigation.

The robustness of implicit cognition

One of the more compelling discoveries about unconscious cognitive processes is that they tend to be more robust than explicit cognitive processes; they typically survive neurological and psychological insults that compromise conscious, explicit processes. Although the vast majority of studies of this property of memory have been carried out only recently, the first hints concerning such a dissociation between the implicit and explicit systems were reported over a century ago by Korsakoff (1889). While working with patients who displayed the syndrome that now carries his name, he noted several cases of memories that, as he characterized them, were "too weak" to enter consciousness but nevertheless seemed to affect his patients' behavior. In one now famous case he gave a patient several mild electric shocks during one session. Later, when he returned carrying the shock apparatus the patient, who apparently had no conscious memory of the earlier episode and who had never seen the device before entering the hospital, exhibited distinct anxiety and accused Korsakoff of coming to give him electric shocks. Some decades later, the French physician Claparède (1911) reproduced Korsakoff's demonstration by surreptitiously sticking an amnesic patient in the hand with a pin upon first greeting her. Later, when Claparède offered his hand, the patient, who had no conscious memory of the earlier meeting, refused to shake it, claiming that people have been known to carry pins around with them. Similar cases have since been reported by many clinicians including Breuer, Janet, Freud, and Prince (see Perry & Laurence, 1984).

The current interest in exploring implicit processes in various patient populations is usually traced back to the work of Milner, Corkin, and their colleagues with H. M., the amnesic patient whose condition followed surgical procedures that involved bilateral excision of the medial temporal region, including removal of the hippocampal gyrus, the amygdala, and two-thirds of the hippocampus. H. M. emerged from the surgery densely amnesic. His anterograde amnesia was so profound that a half hour after eating lunch he could not recall what he had eaten or even if he had eaten at all. Yet H. M. has showed nearly normal abilities in sensorimotor skills such as mirror drawing and tactile maze learning (Milner, Corkin, & Teuber, 1968) and pursuit rotor and bimanual tracking (Corkin, 1968). Improved performance on such tasks clearly requires some memorial residue of the previous experiences. On the other hand, there is no compelling evidence that H. M. shows intact skill learning when the task requires conscious cognitive processes, such as hypothesis testing. On the Tower of Hanoi problem, if care is taken not to prompt him (Gabrieli, Keane, & Corkin, 1987), H. M. shows little or no improvement in performance over trials.

A number of recent examinations of implicit memory and implicit learning in populations of subjects with a variety of psychological and neurological abnormalities reinforced this sense that there was something special about implicit cognition. The unconscious, covert systems are now recognized to be robust in the face of a host of disorders and dysfunctions that seriously impair the operation of the conscious, overt systems.

Not surprisingly, much of the work here has been on patients suffering from *amnesia*. Amnesia is clearly the syndrome of choice since its defining characteristic is an inability to consciously recall and in severe forms to recognize stimuli previously presented. Although the cases studied range across a wide spectrum of disorders and have used a variety of tasks, a common theme has emerged. Behaviors that are dependent on the capacity for conscious, overt recall or recognition of previously presented materials show serious deficits; those that recruit the implicit memorial system show less serious deficits or none at all. These studies cross a remarkable range of findings. What follows is a cursory overview of this literature; in Chapter 3 these findings will be discussed further in the context of the evolutionary model of implicit processes.

Among the more compelling experiments are those in which densely amnesic patients who show chance performance on explicit memory tests, such as recognition and recall, display virtually normal abilities on implicit memory tasks, such as repetition priming or word-stem completion. The first studies here were those of Warrington and Weiskrantz (1968, 1974). Densely amnesic patients were first primed with a set of words. Later, when asked to complete three-letter stems with the first word that came to mind they showed essentially normal tendencies to reply with words that were part of the priming set—despite the fact that they had no explicit memory of those words.

Since then, numerous investigators have replicated and extended these findings with amnesics using a number of related tasks. Cermak, Talbot, Chandler, and Wolbarst (1985) found normal priming using a word identification task in which patients are asked to simply "see" tachistoscopically presented words. Diamond and Rozin (1984) reported a similar priming effect using the stem completion task where patients are asked to complete a word stem with "the first word that comes to mind," and Graf and Schacter in a series of studies (Graf & Schacter, 1985; Schacter & Graf, 1986) reported that amnesics even display context sensitivity in such situations in that they show better stem completion when the test stem is presented in the same context as it was during priming.

Other studies report intact implicit memories in other patient groups with various neuropathologies. One group of disorders here is most revealing. These are the patient populations who, for any of a variety of reasons, suffer from disorders that have essentially "excised" a piece of their perceptual world. In these cases the patients claim to have little or no phenomenological sense of information or knowledge about events of particular kinds yet, under the proper conditions, will show that they have often virtually intact implicit memories of these events.

For example, several research groups have reported that patients with *prosopagnosia* whose effects are often so extreme that patients do not recognize the faces of members of their own family may, nevertheless, show virtually normal implicit facial memory (De Haan, Young, & Newcombe, 1987; Newcombe, Young, & De Haan, 1989; Renault, Signoret, Debruille, Breton, & Bolgert, 1989; Young & De Hann, 1988). In a similar vein, Weiskrantz and his colleagues have examined the phenomenon of *blindsight,* in which patients with scotomas and other visual disorders give clear evidence of covert knowledge of stimuli that have been presented in the damaged sectors in the visual field (Weiskrantz, Warrington, Sanders, & Marshall, 1974; Weiskrantz, 1986). Patients with *hemineglect,* who have seriously impaired ability to attend to or report stimuli in the neglected field, found that they show implicit memories for such events (Volpe, LeDoux, & Gazzaniga, 1979).

Related studies of patients with *acquired dyslexia* and *aphasia* also give support to this general theme. Cases of acquired dyslexia can be particularly revealing since many patients with this disorder still show some reading ability. By careful analysis of the kinds of errors that they make and the manner in which they display their residual abilities, researchers have been shown to implicate the operation of unconscious processes. For example, Shallice and Saffran (1986) and Coslett (1986) both report cases of deeply dyslexic patients whose lesions were such that when asked to read normally they had to resort to letter-by-letter reading strategies. However, when words were presented to these patients at rates too rapid to be decoded they displayed virtually normal lexical knowledge of the presented words on later tests.

The studies with cases of aphasia point to the same conclusion. Andrewsky and Seron (1975) reported implicit processing of grammatical rules in an agrammatic aphasic. They found that, while reading, this patient would either respond correctly or omit a word depending on its grammatical case. When he was confronted with a homophone that could be either a functor or a noun in a sentence, the patient would typically omit it or misread it when its role in that sentence was as a substantive but read it correctly when it was a noun. Blumstein, Milberg, and their colleagues reported in a series of studies (Blumstein, Milberg, &

Shrier, 1982; Milberg & Blumstein, 1981; Milberg, Blumstein, & Dworetzky, 1987) that Wernicke's aphasics had essentially normal patterns of response latencies on lexical decision tasks, despite showing chance performance on explicit knowledge of the meanings of the very same words.

The several experiments that have examined implicit learning in various patient populations report concordant results: implicit acquisition processes show the same robustness as implicit memorial processes. Abrams and Reber (1988) found that psychotics and chronic alcoholics who were incapable of discovering relatively simple letter-to-number rules that were presented as explicit problems nevertheless showed intact ability for implicit learning of the underlying structure of a complex artificial grammar. Nissen and Bullemer (1987) reported that Alzheimer's patients performed normally on an implicit sequence-learning task but poorly on tasks that required reflection and conscious control of cognition. Johnson, Kim, and Risse (1985) showed that amnesics learn to develop preferences for melodies based on Korean melodic patterns that they are unaware they have heard before. In an extended case study, Glisky and her co-workers (Glisky & Schacter, 1989; Glisky, Schacter, & Tulving, 1986) have taught an amnesic patient to operate a computer including being able to carry out data-entry tasks, use disk storage and retrieval procedures, and even write and edit simple programs.

In an interesting experiment, Nissen, Knopman, and Schacter (1987) gave healthy volunteers moderate doses of scopolamine. When compared with a control group, they showed compromised performance on conscious recall and recognition tasks but intact learning and memory on an implicit serial reaction time task.

Finally, a number of experiments carried out using somewhat less "cognitive" tasks have shown similar results. Starr and Phillips (1970) found retention of motor skills in cases of amnesia that were similar to those of Milner and her colleagues with H. M., and Weiskrantz and Warrington (1979) found normal eyelid conditioning in amnesics who had no memory of the earlier conditioning sessions.

Summary

Taken together, these various aspects of the study of the cognitive unconscious have made implicit learning and implicit memory the focus of truly intense empirical and theoretical interest. Cognitive scientists have come to recognize that: (1) there is a good deal of epistemic power in these implicit systems, and (2) any sensible theory of mind is going to have to have in it a rich cognitive unconscious processing system or systems. In the chapters that follow I will pursue several entailments of this work on unconscious cognition. Chapter 2 will present a detailed overview of the work on the acquisition of implicit knowledge. Chapter 3 will develop the evolutionary model in some detail, and Chapter 4 will explore a variety of entailments of this epistemology of the cognitive unconscious.

A personal aside

People often become interested in particular topics because they relate to them in some personal way—at least my clinical friends tell me this is true. I was drawn to the problem of implicit learning simply because that has always been, for me, the most natural way to get a grip on a complex problem. I just never felt comfortable with the overt, sequential struggles that characterized so much of standard learning. *Knowing* something meant being able to function within whatever domain of effective performance that something represented. I think I was a "functionalist" long before I knew that such a perspective on psychological thought existed.

As a result of this stance to matters of education and instruction, I was not a particularly good "standard" student, typically bristling at directed study courses and assigned readings and papers. I found that what seemed for me to be the most satisfactory of "learnings" were those that took place through what we used to call "osmosis," that is, one simply steeped oneself in the material, often in an uncontrolled fashion, and allowed understanding to emerge magically over time. The kind of knowledge that seemed to result was often not easily articulated, and most interesting, the process itself seemed to occur in the absence of efforts to learn what was, in fact, learned. I later discovered that this problem of *tacit knowledge,* its acquisition, and epistemic status had been the focus of considerable philosophical investigation from such learned folk as Wittgenstein, Husserl, Putnam, and most significantly, Michael Polanyi.

Moreover, it seemed that the most basic and critical episodes of acquisition were carried out in this fashion: natural language learning and the acquisition of the skills of socialization. In some very important way, there seemed to me to be a common core to the process through which I was learning some complex topic like literary criticism and how I acquired my natural language and assumed my (reasonably) well-socialized stance within our society. The eventual decision to study the processes that lie at the core of this book was a natural—and made implicitly.

2. Implicit cognition: the data base

This chapter is best thought of as the "empirical" chapter in the sense that the various issues that have been raised by the approach taken to implicit learning will be discussed in the context of the experimental studies designed to explore them.[1] The various methods we have used to examine implicit learning and tacit knowledge will be outlined, and the data base accumulated over the years will be reviewed. The several parallels between our methods and those that various other researchers have used will also be discussed. The focus is on implicit learning and tacit knowledge, and the purpose of the chapter is to draw distinctions between the implicit and nonconscious aspects of cognition and the explicit and conscious. Implicit, nonreflective cognition differs in important ways from explicit, reflective cognition, and these differences lie at the heart of this chapter.

The polarity fallacy

Having said that, a caveat needs to be introduced. It is one thing to have an appreciation of the differences between the implicit and the explicit; it is another entirely to conclude that they are processes of altogether different kinds. We do not want to allow ourselves to be seduced by what we can call, for want of a better name, "the polarity fallacy." That is, we need to be careful not to treat implicit and explicit learning as though they were completely separate and independent processes; they should properly be viewed as interactive components or cooperative processes, processes that are engaged in what Mathews (1991) likes to call a "synergistic" relationship. There is, so far as I am aware, no reason for presuming that there exists a clean boundary between conscious and unconscious processes or a sharp division between implicit and explicit epistemic systems—and no one from Sigmund Freud on has ever argued that there was.

And so, we have a rather tricky issue here. To explicate the distinctions that exist between implicit and explicit cognitive processes, it will often be necessary to present evidence that emphasizes the functional and behavioral differences between them. To convince an audience that the arguments concerning the specialness of implicit learning and tacit knowledge are sound, it becomes incumbent upon the proponent of the theory to sharpen differences and soft-pedal similarities. This is unfortunate but, given the nature of the give and take of academic and scientific discourse, unavoidable.

There are, as will be outlined in this chapter, many reasons, empirical and theoretical, for drawing distinctions between implicit learning and explicit learn-

[1] The material in this chapter is derived in part from two previously published papers (Reber, 1989a, 1989b).

ing and between tacit knowledge and knowledge that can be easily explicated. In later chapters, specifically Chapter 3 and to some extent Chapter 4, evolutionary arguments will be put forward to draw distinctions between the phylogenetic histories of these systems. However, there will not be any evidence to show that these domains of cognitive function should be treated as utterly separate.

In fact, there are no reasons, empirical or theoretical, for assuming that there is any well-defined cut-point or threshold separating the two at some point along a continuum. Implicit and explicit systems should properly be viewed as complementary and cooperative functional systems that act to provide us with information about the world within which we function. As will become clear, virtually every interesting experiment that has been done on implicit learning contains some explicit elements—and the reverse is probably true as well. The interesting psychological issues have to do with balance and not with exclusionary clauses.

Our science has had some dreadful experiences with issues like this one. Virtually every time that a continuum of processing or performing is discovered we seem to fall into the fallacy. Once again, a simple perusal of standard texts and journals proves somewhat depressing. People are classified as internal or external in terms of locus of control, masculine or feminine in terms of sex roles, intrinsic or extrinsic in their motivation, conformers or nonconformers in social behavior; they are introverts or extroverts, have Type A or Type B personalities; knowledge is either episodic or semantic, memories are propositional or imaginal, reasoning is inductive or deductive; we are sharpeners or levelers, and so on.

Interestingly, in situations where the phenomena were introduced by labeling the continuum independent of the extremes, the problems were ameliorated. When the polar terms are not used, the continual nature of the dimension tends to be conceptually maintained. Psychological dimensions like need-achievement, manifest anxiety, depression, hypnotizability, learnability, and so forth, tend to be dealt with in this fashion. References are made to people being at some point along the continuum rather than representing them in terms that affix them to one of the poles. In such situations it is easier to keep clear the fact that the poles are indeed poles and not points on psychologically distinct dimensions.

The point of all this is to recognize that some of the disputes over the implicit or explicit characteristics of a particular process are fueled by our affection for the polarity fallacy and are not nearly so disputatious as they first appear. This will become apparent when we examine several of the issues raised by the experiments reviewed in this chapter.

On the primacy of the implicit

Various researchers have argued that there really is nothing particularly interesting about many of the studies of implicit learning (e.g., Brody, 1989; Dulany, Carlson, & Dewey, 1984, 1985). One of the more significant features, the nonverbalizable properties of the knowledge can, it is argued, be viewed as little other than a failure to probe effectively the subjects' ability to explicate their knowledge. In short, implicit learning is implicit only to the extent that experimenters have failed to make it explicit. If there is evidence that subjects have conscious knowledge then the learning cannot have been implicit.

This "nothing special" line of argument is a direct entailment of taking the stance that consciousness is primary and that the default position should be that conscious processes lie at the heart of human cognition. From this perspective, unconscious, implicit functions are dealt with derivatively and virtually all interesting cognitive functions are to be seen as dependent on conscious processes. Implicit learning, here, is merely an illusion resulting from the failure of the experimenters to ask their subjects the right questions; action independent of awareness is due to the automatization of behaviors whose acquisition was dependent on awareness (Dulany et al., 1984; Brody, 1989).

I will have much to say about this line of argument in this and later chapters. It should be quite clear that the position I take throughout is that of the *primacy of the implicit*. As will be argued below, other things being equal, implicit learning is the default mode for the acquisition of complex information about the environment. The full development of this perspective will be found in Chapter 3 where the phylogenetic and ontogenetic considerations driving this argument are developed.

On functionalism

The overall perspective on implicit processes we have typically taken is that of the old warhorse, *functionalism* (see Reber, 1989a; Reber & Allen, 1978; Reber & Lewis, 1977). It is a position with often surprising explanatory force and the virtues of deep roots in the thinking of Dewey, James, and Peirce. It is also a position that helps keep one honest when it comes to the aforementioned fallacy. From the functionalist point of view, implicit and explicit modes of operation are, under normal conditions, both "available" as viable procedures for acquiring information about the world. The circumstances of acquisition, the constraints of the task, the demands of the experimental setting, the *Einstellung,* to use Külpe's term, dictate the extent to which each is engaged. Under nonexceptional circumstances, complex human behavior is a delicate blend of the implicit and the explicit, the conscious and the unconscious.

In various special situations where tight experimental controls have been applied or various kinds of special circumstances pertain, one may see a shift in the balance even, perhaps, to the extent that one mode (or the other) virtually dominates performance. These, of course, are the very circumstances one attempts to establish in order to examine one mode (or the other).

Some assumptions

The remainder of this chapter will introduce various experimental settings where implicit induction processes achieve a considerable measure of dominance over the explicit. In later chapters, specialized circumstances will be reviewed that also seem to differentially recruit implicit and explicit modes, such as childhood, some psychiatric syndromes, and a variety of neurological disorders. The empirical studies that follow have formed the heart of our continuing research program. They were carried out with a number of presumptions about implicit processes that have functioned as our conceptual framework:

1. That the processes of implicit induction are general and universal.
2. That implicit learning is a foundation process that operates as a base for the development of tacit knowledge of various kinds.
3. That there are no a priori reasons for making assumptions about biological determinants of specific kinds.
4. That the properties of the tacit knowledge base that develops from implicit learning are reflective of the structure inherent in the stimulus environment.
5. That implicit acquisition is the default mode and the one normally adopted.
6. That when procedures are modified, functional considerations will dictate the degree to which implicit and explicit processes will be recruited.

Taken together, this cluster of assumptions represents a fairly strong form of classical representational realism. Philosophers have had a fine old time trashing realism, accusing it of, among other sins, being hopelessly naive. While I will acknowledge that it is just a tad unsophisticated philosophically, I believe that this unabashedly critical stance is just a tad misplaced. As the material in this chapter is developed, I hope that it will gradually become clear that representational realism is a position that has considerably greater explanatory power than has generally been appreciated. There are fairly good reasons for concluding that the structure that is out there "in the world" is the structure that ends up "in the head" (see Reber, 1991).

Experimental procedures

Research in implicit learning is properly carried out using arbitrary stimulus domains with complex, rule-governed, idiosyncratic structures. Ideally, in order to obtain insight into a process such as implicit learning, it is essential to present the learner with a stimulus domain that has the following properties:

1. The stimuli need to be novel, the rule system that underlies them cannot be something that is already within the subjects' sphere of knowledge—this, of course, is an important heuristic first used over a century ago by Ebbinghaus.
2. The rule system that characterizes the stimulus domain needs to be complex. If subjects are able to "crack the code," as it were, by simple testing of consciously held hypotheses, implicit learning will not be seen.
3. The stimuli should be meaningless and emotionally neutral, or as devoid of meaning and affect as one can make them. The point here is to diminish extraneous aspects, particularly aspects that might have differential effects on individual subjects.

4. The stimuli should be synthetic and arbitrary. If our assumptions about implicit learning are correct, it should appear when learning about virtually any structured stimulus domain and the use of the synthetic and the arbitrary gives additional force to the argument.

As will become apparent, not all of the investigations into implicit learning have used stimulus domains with these properties. Berry, Broadbent, and their colleagues, for example, have made extensive use of a class of experiments involving various "control" systems. In these settings, subjects have to learn to function in complex environments to control such factors as the production of materials in a hypothetical manufacturing plant or to modify the affective responses of some fantasy person (see here, Berry & Broadbent, 1984, 1987; Broadbent & Aston, 1978; Broadbent, FitzGerald, & Broadbent, 1986). Within these much less obscure stimulus worlds, implicit acquisition of the rule system has been reliably reported. Subjects acquire knowledge of rule systems that define the environment enabling them to gain control over the stimulus domain, but are largely unable to explicate the rules they use.

Mathews and his co-workers (Mathews, Buss, Chinn, & Stanley, 1988; Mathews, Buss, Stanley, Blanchard-Fields, Cho, & Druham, 1989; Stanley, Mathews, Buss, & Kotler-Cope, 1989) have employed a number of tasks including concept induction using Chinese-like ideograms, Broadbent-type control tasks, and artificial grammars of the kind we have used. Lewicki and his colleagues have used a wide variety of different stimulus settings from the highly arbitrary (a matrix-scanning task using patterns of stimuli that followed complex positional rules; Lewicki, 1986b, Lewicki, Czyzewska, & Hoffman, 1987; Lewicki, Hill, & Bizot, 1988) to the mundane (photographs of young adults accompanied by personality descriptions; Hill, Lewicki, Czyzweska, & Boss, 1989; Lewicki, 1986a; Lewicki, Hill, & Sasaki, 1989). Nissen and her colleagues (Nissen & Bullemer, 1987; Nissen, Knopman, & Schacter, 1987; Willingham, Nissen, & Bullemer, 1989) have used light sequences that follow complex patterns, which subjects have to learn. Howard and his co-workers (Howard & Ballas, 1980, 1982; Howard, Mutter, & Howard, 1992) have used various environments including artificial languages with and without semantic components and Nissen-like patterns. Finally, several research groups have used complex sequences of stimulus events in which the sequences are structured according to complex rules. These experiments range from those that use relatively simple, short sequences (such as those employed by Cohen, Ivry, & Keele, 1990) to those that are most complex and based on the probabilistic structure of artificial grammars (such as those employed by Cleeremans & McClelland, 1991). In all of these cases, unconscious apprehension of covariations between elements of the stimulus set were observed.

These, and still other techniques that have been used with a variety of subject populations, will be discussed in more detail below. First, a review of our work is in order.

We have chosen to work with two procedures that we have found to be extremely useful: artificial-grammar learning and probability learning. The former is, by now, well known in the literature and has been used by many; the latter, once a common method for studying conditioning-like processes, is somewhat obscure in the context of implicit learning. However, as will become clear, it is an exceedingly sensitive technique that has provided some intriguing data.

Grammar learning

Figure 2.1 shows one of the first synthetic grammars used along with some examples of the basic types of "sentences" it can generate. This artificial grammar (AG) was first used in the earliest studies of implicit learning, which were carried out in the early 1960s (Reber, 1965). Several other grammars of varying complexity have also been used by us and others; examples are given in Figure 2.2. All are Markovian systems derived from a simpler system that formed the basis of George Miller's *Project Grammarama* (see Miller, 1967).

One of the reasons for choosing at the outset systems like these to generate stimulus items is that there is a well-developed branch of finite mathematics that underlies them. Since, as mentioned above, it is important to use stimulus domains that are arbitrary, the proper course seemed to initiate the investigations with stimuli whose formal properties were relatively well known.

For those who are interested, an introduction to finite-state grammars can be found in a paper by Chomsky and Miller (1958). As they show, it is relatively easy to calculate the complexity of such systems using standard information theory algorithms, and it is possible to state specifically just how much *information*, in the technical sense, each string in a Markovian grammar carries. Such measures turned out not to be of much use in psychological science; information, measured in bits, has little relationship to subject performance, at least within the ranges we have looked at. Nevertheless, it seemed prudent to opt for well-known formal systems at the start, and interestingly, the grammar learning paradigm has proven extremely powerful and generally independent of these formalist considerations.

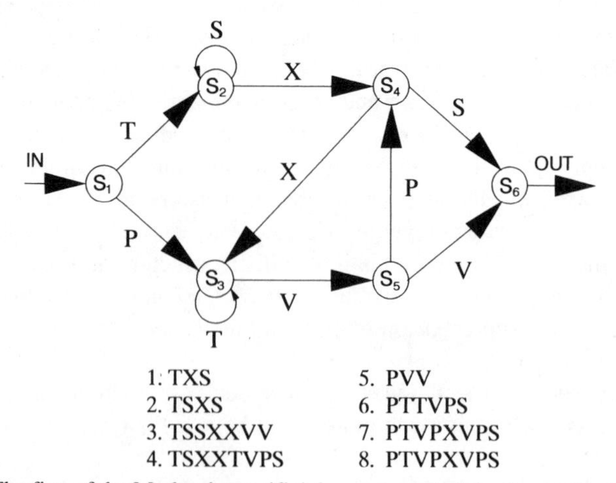

Fig. 2.1. The first of the Markovian artificial grammars (originally used in Reber, 1965) with several examples of grammatical strings that it generates.

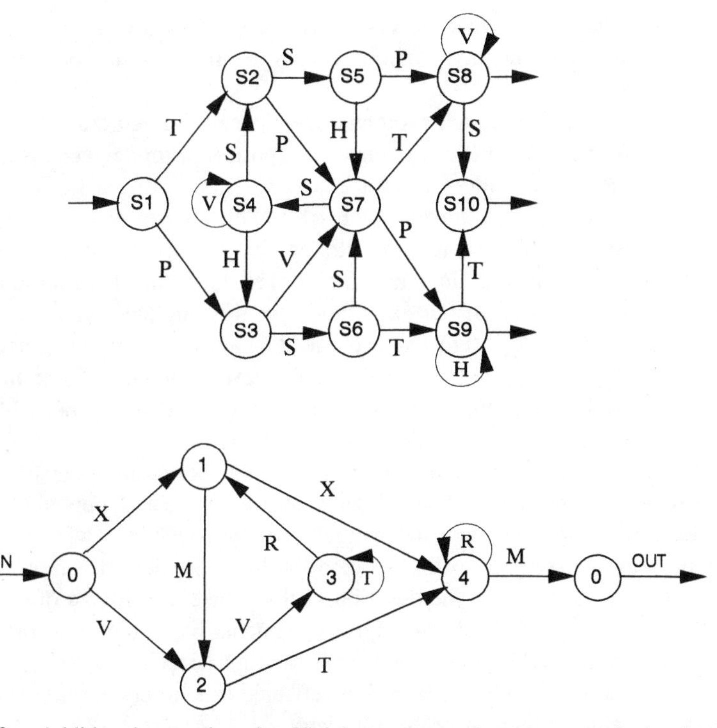

Fig. 2.2. Additional examples of artificial grammars of varying complexity that have been used in a variety of studies.

Although there have been many variations on a theme here, the basic procedure used in the grammar-learning studies is straightforward and has two essential components. There is an acquisition phase during which subjects acquire knowledge of the rules of the grammar and a testing phase during which some assessment of the subjects' knowledge is made. Additional details will be provided later when needed to explicate particular experiments.

Several points, however, need to be kept clear about these synthetic languages and the manner in which they have been used to examine implicit, unconscious cognitive processes. First, they are complex systems, "too complex to be learned in an afternoon in the laboratory," as Miller put it in 1967. Miller originally saw this as a liability, which it is if one wishes to examine explicit concept learning—the original intent of his *Project Grammarama*. This complexity, however, is a virtue in the current context since a rich and complex stimulus domain is a prerequisite for the occurrence of implicit learning. If the system in use is too simple, if the code can be broken by conscious effort, then one will not see implicit processes.

Second, the fact that the grammars used here are finite-state, Markovian systems means that they generate strings of symbols in a left-to-right, nonhierarchical fashion. This property should not be taken as reflecting any prejudices about the structural underpinnings of natural languages or their acquisition. The decision to use finite-state grammars was motivated by the various reasons already discussed quite independent of theoretical issues in linguistics or natural language learning.

From time to time, research programs have been mounted that have tried to use such "miniature linguistic systems" as settings within which to examine natural language acquisition (see, for example, Braine, 1971; Braine, Brody, Brooks, Sudhalter, Ross, Catalano, & Fisch, 1990; Moeser & Bregman, 1972; Moeser & Olson, 1974). Our research strategy has been quite far removed from these. I view the synthetic language as a convenient forum for examining implicit cognitive processes. If our explorations of this implicit domain of mentation turn out to provide insights into the acquisition of natural languages, that would be most pleasing. Any insights here, however, will come from an understanding of the learning process independent of the fact that some of the data were gathered using artificial grammars.

Moreover, as will become clear, there is nothing special about these stimulus generators in any interesting psychological sense. The basic components of implicit learning have been seen to emerge, as mentioned, in a wide variety of different empirical settings using a range of different stimulus environments. This is an important point. The question of the generalizability of our reported findings from these artificial language experiments has often been discussed, occasionally loudly. The point that I wish to emphasize is that the broad data picture that emerges from these studies is reflected in a variety of other empirical settings, including studies of probability learning.

Probability learning

The format we adopted here departs noticeably from the traditional two-choice procedure in which each trial consists of a ready signal, a prediction response, and an outcome event. Because our conceptualization of this experimental paradigm is a bit radical, it will help to start with a little background.

The "standard" probability-learning (PL) experiment has a long and checkered history. It was developed over a half century ago by Lloyd Humphreys (1939), who introduced it as an analog to a conditioning experiment. In the early experiments, each trial consisted of a ready signal, a prediction response in which the subject indicated whether or not a "reinforcing" event (a light) would occur, and an outcome that was the occurrence or nonoccurrence of that light. Humphreys's view of the task as an analog to a conditioning experiment was based on the assumption that the subject was making a covert verbal gesture, as it were, about the outcome of each trial, and this response was being reinforced by the actual outcome. By varying the likelihood of the outcome light, parameters like the probability of reinforcement could, in principle, be easily explored.

Over a decade later, William Estes and his co-workers modified the general Humphreys's technique in simple but important ways. Estes saw the PL experiment as a procedure within which to test various entailments of his statistical learning theory (see Estes, 1959 and 1964, for overviews of this orientation). The major methodological adjustment Estes made was to have separate, specific responses that the subject could make as predictions. In other words, rather than have a single button, which was either pressed or not, Estes's experiments had two (or more) buttons from which the subject had to choose and, of course, two (or more) corresponding outcome events.

With this simple modification, the procedure provided a clean experimental setting in which to subject the then newly developed mathematical theories of learning to some very strong experimental tests. The probability learning procedure quickly became a classic. It contained all of the essentials for an associative acquisition process to occur: a well-defined stimulus that set the stage for a response, a fixed number (usually two) of unambiguous responses among which the subject had to choose (denoted A_1 and A_2), and a similar number of reinforcing outcome events (E_1 and E_2)—the response that occurred specified, and presumably reinforced, the corresponding response. By controlling the probabilities of the outcome events, powerful control over a subject's behavior could be established.

As is reflected in the now-accepted name of the experimental paradigm, it was quickly noted that subject's prediction responses tended to match (more or less) the probability of the several outcome events. If there were two events, E_1 and E_2, and they occurred with probabilities of, say, .80 and .20 respectively, then subjects' average response rates for the two responses, A_1 and A_2, over trials, typically came to approximate .80 and .20. Hence, the subjects displayed *probability learning*. Estes's statistical learning theory (SLT) gave what appeared at first to be a rather remarkably accurate account of the data (see Estes & Straughan, 1954). In fact, at that time there was no other theoretical account that would have predicted anything other than "maximizing." The prediction of the then "standard learning theory" was that, at asymptote, the likelihood of the A_1 response would approximate 1.0 in any situation where the E_1 event had a probability of occurrence greater than .50.

However, with the intense empirical scrutiny that success like this typically spawns, a number of difficulties with the standard model emerged. Subjects began to show a lamentable tendency to respond in ways counter to the predictions of SLT; they acted, of all things, like they had memories. The early versions of statistical learning theory, however, were based on a simple associative process with independence of path assumptions; there was no provision for memorial functions.

A common finding was that as the length of a run of a particular event increased, subjects showed an increasing tendency to predict the other event. It was as though they were saying, "the right light has come on so many times in a row now that the left is due." Jarvik (1951) called this effect *negative recency;* it is also known as the *gambler's fallacy* and can be readily observed in any gambling casino. According to Estes's theory, however, increasing runs of event E_1 should produce exactly the opposite, subjects should show an increasing likelihood of making the A_1 response. Negative recency showed that subjects were sensitive to the sequence of earlier events and, as Jarvik (1951) had noted, it only took runs of lengths 2 or 3 to produce statistically detectable effects.

Various other studies (e.g., Anderson, 1960) showed that subjects also had an extreme sensitivity to single and double alternation sequences of events. The probability of an A_1 response was found to be very high following sequences of *E*s, like 121212 and 11221122. Indeed, as with the recency effects, even a simple sequence like 212 could provoke a statistically detectable increase in the likelihood of subjects making an A_1 response.

A number of theoretical adjustments were made in an attempt to account for findings such as these. Burke and Estes (1957), for example, developed statistical learning theory models that had a memorial component. These models assumed that subjects kept track of some number of earlier events and used the patterns of events as the basis for their prediction responses. Here, the subject was conceptualized as having a limited, short-term memory system, which represented the earlier sequence of trials as a pattern, and each pattern was characterized as being conditioned to particular prediction responses. This so-called "pattern model" was an advance in that it circumvented the path-independent assumptions of the earlier model and gave a better account of the data. However, as was quickly shown, there were still serious problems in accounting for the observed data.

Specifically, there was still no ready explanation, even within the confines of these more sophisticated models, why some patterns of events should have greater psychological salience than others. That is, the models predicted that a sequence of events like 122112 should function as a component cue and become conditioned with the same number of presentations as the sequence 111111 or 121212. This prediction was not supported by fine-grain analyses of the data. Moreover, the new models still could not deal with the negative recency effect, and this turned out to be critical.

Gradually it became clear that the probability-learning experiment was not the uncluttered setting it was first thought to be. Several researchers (Goodnow, 1955; Feldman, 1963) put forward intriguing cognitive theories of PL that gave a rather distinct slant on the experiment. Subjects were no longer seen as the passive recipients of response-outcome pairings as in the statistical theories; they were characterized as intensely active participants in the experiment. Subjects were assumed to be testing hypotheses about the patterns reflected by the sequence of events. If the statistical theories were essentially models of data; these were models of process.

Feldman's approach was particularly interesting in that it made quite a few assumptions about the subjects' mental processes and tried to trace the data from the experiments back to these mental processes via the subjects' verbal reports of the hypotheses they used in making the predictions. Feldman ran subjects with an open microphone on the table in front of them and encouraged them to give a running account of their reasons for making each prediction response. He found, among other effects, that subjects had a pretty good memory for recent past events. He also found that they told him that they were (in so many words) committing the gambler's fallacy and showing extreme sensitivity to single and double alternations.

They also showed remarkable tenacity in their role as problem solvers, for that is how they reported they saw the experiment. Despite the fact that the event sequences were Bernoulli in nature, the typical subject was convinced that patterns were present and persisted in searching for them. Prediction failures were, on more than a few occasions, accompanied by accusations that the experimenter had changed the rule. From this stance, the standard probability learning experiment had changed radically. It was no longer an experiment on passive conditioning but had become a study in explicit, overt, rule learning.

When I first approached the PL paradigm (Reber, 1967b), I saw it as neither of these. Not surprisingly perhaps, I viewed it as containing many properties that lent it to an examination of implicit learning. In an extended series of studies with Richard Millward, we explored this characterization using, as I stated at the outset of this section, a format that departed noticeably from the traditional procedure.[2]

The procedure we adopted to explore implicit processes derives from the proposition that the essential nature of a PL experiment has little or nothing to do with the explicit learning of probabilities of events and, with a few modifications, can be shown to have equally little or nothing to do with explicit rule learning. Instead, what has passed in the literature for probability learning is actually a much more subtle process in which subjects learn implicitly about the stochastic structure of an event sequence to which they have been exposed. In the course of making prediction responses, they mimic the structure of the event sequence and thereby generate a sequence of responses, one byproduct of which is an approximate matching of the probabilities of events, that is, "probability learning."

Accordingly, the procedure was modified as follows (Reber, 1967b; Reber & Millward, 1968). The subject begins an experimental session simply by observing the occurrence of a sequence of rapidly presented events. There is no ready signal and the subject makes no prediction responses. In this situation, a passively observed event turns out to be functionally equivalent to a traditional trial (Reber & Millward, 1968) and a learning session consisting of two or three minutes of observing events at a rate of two per second is sufficient to put a subject at an asymptotic rate of responding. That is, subsequent prediction responses made by subjects who have had this learning experience show all of the characteristics of ordinary subjects who have had an equivalent number of traditional trials. We dubbed this procedure the *instant asymptote* technique.

The typical experiment with this modified PL procedure consists of an acquisition phase during which subjects observe event sequences, which may have

[2] For those readers who choose to track down the original articles being reviewed here, be forewarned. It will difficult to see in them any of the traces of implicit learning theory; these articles were victims of the *zeitgeist*. At the time, the PL experiment was being used primarily to test various entailments of statistical learning theory. Pressure from editors, reviewers, and the not-so-subtle prejudices of the community of scholars of which we are all a part conspired to produce what, in retrospect, seems to me now to be a series of dry, even boring papers dripping with equations and cold logic. Part of the reason for going into the kind of detail that I do here is to redress this situation.

any of a variety of stochastic structures, and a testing phase where subjects make prediction responses in which they reflect the knowledge acquired. In the very first studies, we used only Bernoulli sequences of events so the term "knowledge" should be used here with appropriate caveats. We used these unstructured event sequences simply because we were concerned with establishing the "deep" equivalence between the instant asymptote technique and the processes that had been intensely studied by so many others using the more typical ready-signal → prediction-response → outcome-event trials. On the face of it, it seemed mildly counterintuitive that 3 minutes of observation could have the same impact on subsequent behavior as 45 to 60 minutes of ordinary PL training.

Once the equivalencies were strongly established (see Reber & Millward, 1968), we began introducing variations in which the event sequences obeyed various stochastic rules and subjects' response sequences were examined for the extent to which they reflected these patterns. As with the grammar-learning studies, there are many variations on this basic theme; they will be introduced as needed.

Note that despite the many superficial differences between the probability-learning paradigm and the grammar-learning experiment, there are two essential commonalities. First, in both cases the subject is initially confronted with a stimulus environment about which knowledge must be acquired in order to respond effectively during the testing session. Second, in both cases stimulus domains are being used whose underlying structures are not part of nor even remotely resemble the epistemic contents of the typical subjects' long-term memories. These points are fundamental; the whole purpose of examining implicit learning in the laboratory is to develop understanding of how rich and complex knowledge is initially obtained independent of overt, conscious strategies for its acquisition. This process is ubiquitous in human experience; yet, as a focus for psychological inquiry, it has been largely unrecognized.

What follows is an overview of several dozen experiments using these two procedures, which were carried out over the past several decades. They are presented in the form of basic issues to which the literature speaks. This material is integrated with the growing body of literature on unconscious, nonreflective, implicit processes. I will then be able to use the material in this chapter as a kind of empirical "launching pad" from which to theorize about how such cognitive systems could evolve and how they fit into the contemporary struggles with a number of classic problems in pure and applied psychology.

Finally, note that the work in our laboratory that is the focus of this overview, beginning with the very first studies (Reber, 1965, 1967a), was carried out with a particular research strategy very much in mind: use a limited number of techniques to examine a wide variety of effects. The virtue of this approach is that by developing a few techniques and building a robust data base, one can explore a large number of issues and not be terribly concerned about the vagaries that get introduced with alternative procedures. Given that the problems of implicit learning and tacit knowledge can be explored using these two procedures, this heuristic says that they should be used in as many circumstances as makes scientific sense. Biological fans of the *E. coli* microorganism will recognize the strategy.

There is, of course, an alternative strategy: examine these unconscious cognitive and perceptual processes across as broad a range of experimental environments as possible. The virtue of this strategy is that one is not likely to be taken in by idiosyncratic properties of particular procedures; generalizations come easier. If implicit learning is real, it may be legitimately argued, it should emerge in contexts conceptually remote from synthetic grammars and structured event sequences. Ideally, of course, both strategies should be carried out. As will become clear, those whose research programs have taken the latter tack, such as Donald Broadbent's, Paul Lewicki's, and Robert Mathews's, have typically produced data that are congruent and complementary. The generalizations implicit in such congruency are, of course, critical if anything serious is to come of this endeavor.

Empirical studies of implicit learning

On the exploitation of structure

When people are presented with an environment that is structured, they learn to exploit that structure. That is, they come to use their knowledge of the structure to behave in a relevant fashion in its presence. This proposition seems noncontroversial as a generalization about human cognition. Indeed, it seems to be almost a signature of our humanity. In one form or another, it lies at the core of any number of approaches to various problem areas in psychology such as perception (Gibson, 1966, 1979; Mace, 1974) and decision making and information processing (Garner, 1974; Hasher & Zacks, 1984). It may also be detected as underlying, in a broad sense, any of a number of general theoretical analyses such as Anderson's (1983) production systems; Neisser's (1976) ecological approach; Nelson's (1986) and Schlesinger's (1982) models of natural language acquisition; Lewicki's (1986a) analysis of socialization; Holland, Holyoak, Nisbett, and Thagard's (1986) model of induction; and Rumelhart and McClelland's (1986) connectionist system.

However, despite the widespread recognition that this capacity is so basic, there has been surprisingly little "pure" work done on it. By "pure" I mean research in which the structures to be learned, elaborated, induced, and used are novel and one can watch, as it were, the exploitational capacity emerge uncontaminated by previous knowledge. In an important way, providing just such a demonstration is the first obligation that a defender of implicit learning theory assumes.

GRAMMAR-LEARNING STUDIES. The very first study with artificial grammars (Reber, 1965, 1967b) showed that subjects became increasingly sensitive to the constraints of the grammar they were working with simply by exposure to exemplary strings. In this experiment subjects were purposely not informed that they were working with rule-governed stimuli. The experiment was pitched as a study in human memory, and subjects were requested to memorize strings of letters printed out on cards, four strings to a card.

Memorizing 4 strings of letters generated by a Markovian system of the kind in Figure 2.1 is no mean task. It is possible to get a feeling for what these stimuli are like by perusing Table 2.1, which lists 20 strings of the kind used here (under "Learning Stimuli"). Notice that in this study maximum string length was 8; this is typical of these studies although in some cases lengths as great as 11 have been used (Millward, 1981; Servan-Schreiber & Anderson, 1990). Being very pragmatic here, our main concern dictating item length has been to ensure that one can generate a sufficient number of stimuli for the needs of the experiment. The grammar in Figure 2.1 has a "vocabulary" of 43 distinct strings of lengths 3 through 8, more than enough for the typical study. However, as we discovered

Table 2.1 An example of the stimuli used in a typical AG study. The 20 Learning Stimuli are used in the acquisition phase, usually by asking subjects to memorize them. The Testing Stimuli are used in the well-formedness task where subjects are asked to determine the grammatical status of each string when presented. These strings are generated by the AG shown in Figure 2.1.

Learning Stimuli	Testing Stimuli	
1. PVPXVPS	*1. PTTTVPVS	*26. SVPXTVV
2. TSSXXVPS	*2. PVTVV	27. PVPXTTVV
3. TSXS	*3. TSSXXVSS	28. PTTVPXVV
4. PVV	*4. TTVV	29. TSXXTVPS
5. TSSSXXVV	5. PTTTTVPS	30. TXXTVV
6. PTVPXVV	6. PVV	31. TSSSSXS
7. TXXVPXVV	*7. PTTPS	*32. TSXXPV
8. PTTVV	8. TXXTTVPS	33. TPVV
9. TSXXTVPS	9. TSXXTTVV	*34. TXPV
10. TXXTVPS	*10. PVXPVXPX	*35. TPTXS
11. PTVPS	*11. XXSVT	36. PVPXTVPS
12. TXS	12. TSSXXTVV	*37. PTVPXVSP
13. TSXXTVV	13. TXS	38. PVPXVV
14. PVPXTVPS	*14. TXXVX	39. PTVPXVPS
15. TXXTTTVV	*15. PTTTVT	*40. SXXVPS
16. PTTTVPS	16. TSXXVPS	41. TXXVV
17. TSSSXS	17. PTTTVV	*42. PVTTTVV
18. TSSXXVV	*18. TXV	43. TSSXXVPS
19. PVPXVV	19. PTTVPS	*44. PTVVVV
20. TXTVPS	20. TXXTTVV	*45. VSTXVVS
	*21. PSXS	46. TSXXVV
	*22. PTVPPPS	*47. TXXTVPT
	23. PTTTTTVV	48. PVPS
	*24. TXVPS	*49. PXPVXVTT
	25. TSSXS	*50. VPXTVV

*Indicates a nongrammatical string

later, the string length parameter had some unexpected effects, which are discussed below.

Committing to memory four strings with lengths 3 through 8 is, as mentioned, not easy. However, with practice subjects become increasingly adept at this task. Figure 2.3 gives the data from Reber (1967a). There it can be seen that subjects who worked with these grammatical strings (*G*) made an average of nearly 18 errors before being able to reproduce all four strings of the first set correctly; by the seventh set they made, on average, fewer than 3. The subjects who worked with random strings made up of the same letters but with order dictated by a random number table (*R*) showed a distinctly different pattern. They, too, committed nearly 18 errors at the outset, but since there was no structure to exploit, the curve flattens out at about 8. The improvement from Set 1 to Set 2 is a kind of "learning to learn" effect that occurs independent of structure.

Moreover, following this neutral learning task with the rule-governed strings, subjects were able to use what they had apprehended of the rules of the grammar to discriminate new strings that conformed to the grammatical constraints from those that violated one or more of the rules of the grammar (see Table 2.1 for examples of these strings). In simplest terms, these subjects can be said to be exploiting the structure inherent in the stimulus display. This basic finding, in experimental settings like this one, has proven to be a robust one. It has been replicated in our laboratory on innumerable occasions and by Brooks (1978);

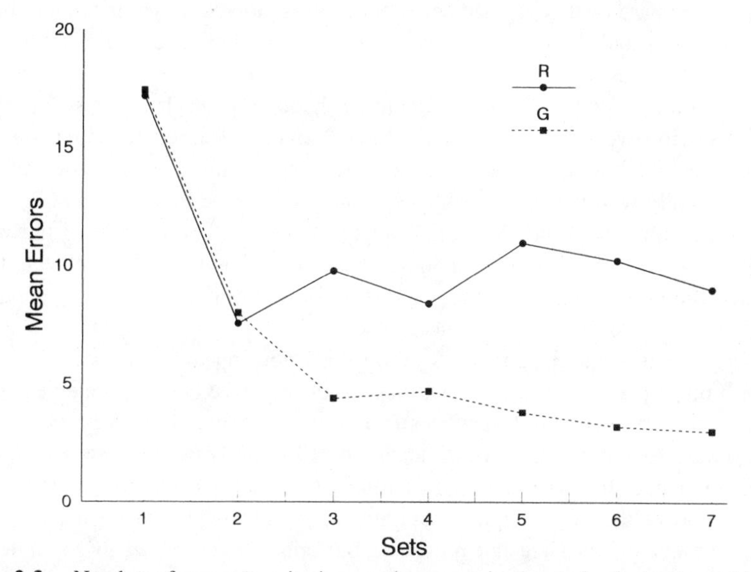

Fig. 2.3. Number of errors to criterion on the memorization task. G denotes subjects who worked with structured, "grammatical" strings; R denotes those who worked with unstructured, "random" strings. This figure is adapted from "Implicit learning of artificial grammars" by A. S. Reber, 1967, *Journal of Verbal Learning and Verbal Behavior, 6*, p. 325. Copyright 1967 by Academic Press.

Dulany et al. (1984); Fried and Holyoak (1984); Howard and Ballas (1980); Mathews, Buss, Stanley, Blanchard-Fields, Cho, and Druhan (1989); Millward (1981); Morgan and Newport (1981); and Servan-Schreiber and Anderson (1990), among others. It is, in fact, so replicable that it has been included as a laboratory demonstration in the supplement to a major introductory psychology text (Gleitman, 1981).

PROBABILITY LEARNING STUDIES. Several of the PL studies have also yielded analogues of this process. Reber and Millward (1971) found evidence that subjects can accurately anticipate the changing probabilities of events even when the anticipatory response requires an integration of information across 50 preceding events. In this particular case, the probability of each individual event on any Trial *n* was systematically increased and decreased as *n* moved through a period of 50 trials. Subjects were first given 1000 instant asymptote trials with this "sawtooth" event sequence at the rate of 2 per second and then requested to predict successive events prior to their occurrence. The experiment was run over three days with different patterns of outcome events on the testing phase used on each day.

The instant asymptote phase of all of these PL experiments consists essentially of watching two lights flash on and off rapidly. In this study the sense one gets is of a shifting "density" of lights. The left-hand light, for example, gradually increases in frequency up to some vague maximum and then gradually becomes less likely, while the right-hand light shows the complementary pattern. During the testing phase of this experiment everything is slowed down; the signal light is introduced, and the subjects have to make a prediction response before the event light comes on.

Under these conditions, subjects begin by shadowing the changing event probabilities. However, with experience they ultimately learned to anticipate the shifts in the likelihood of events, so that their predictions responses rose and fell coincidentally with the actual event sequences. As can be seen in Figure 2.4, by the nineteenth and twentieth 50-trial blocks, the peaks and troughs of the subjects' response curve and the outcome event curve coincide. The subjects had learned the underlying structure of the stimulus environment and were capable of exploiting it to direct their choices.

Note that this anticipatory responding does not appear right away, it takes quite a bit of practice with the wave-like event sequence before it emerges. The nineteenth and twentieth trial blocks in Figure 2.4 come after 38 previous 50-trial blocks (counting the learning phase) or 1900 trials of experience with the event sequence. It is also not terribly robust behavior and requires maintaining the support of the event sequence. Figure 2.5 shows what happened on Day 2 when the event lights are not presented but subjects are asked to continue to respond "as though they were there." Here, the cycling damps out quite quickly. Figure 2.6 shows the effects of the final removal of the wave form by introducing simple Bernoulli sequences. On these trials, subjects reverted to "standard" PL-type behavior with a slight overshooting of the probability of the more frequent event.

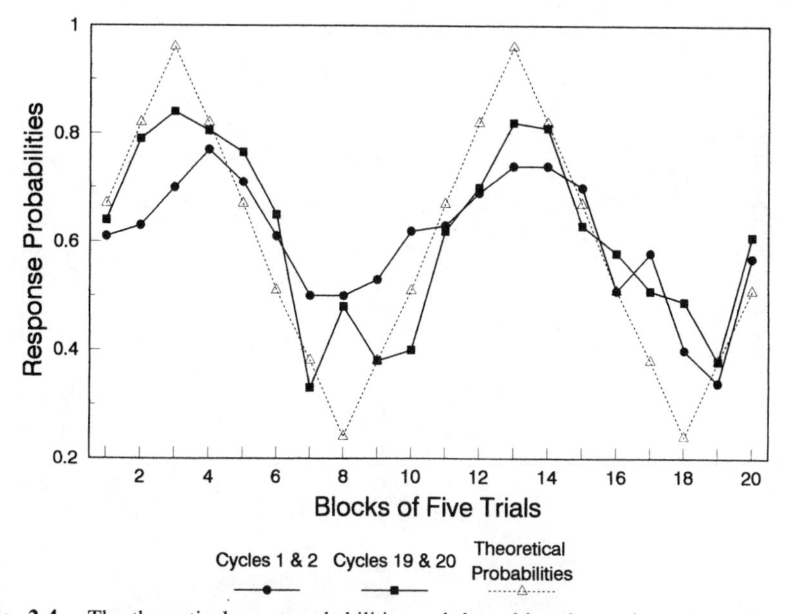

Fig. 2.4. The theoretical event probabilities and the subjects' actual response proportions in blocks of five trials for the first two cycles and the 19th and 20th cycles of training on a "sawtooth" event sequence. This figure is adapted from "Event tracking in probability learning" by A. S. Reber and R. B. Millward, 1971, *American Journal of Psychology, 84,* p. 89. Copyright 1971 by the American Journal of Psychology.

Nevertheless, the fact that the effect appears at all is rather impressive. For subjects to respond as they did requires the integration of a subtle wave pattern with a 50-trial phase. Indeed, if you wish to push at this point from another direction, there is another way to think about it. It is possible to regard each run through a cycle as a single trial; in this case, the subjects are learning about the structure of the event sequence in a "mere" 35 or 40 trials. It should be noted that in several of Lewicki's experiments using similarly complex stimulus sequences, the evidence for implicit learning also did not emerge until there had been considerable practice (see below, p. 43). And, just in passing, it is worth pointing out that none of the conditioning theories of PL can handle this finding. They all predict that subjects' response probabilities will follow the shifting probabilities of events, not anticipate them.

Millward and Reber (1972) reported an even more impressive ability of subjects to exploit stochastic structures using event sequences with short- and long-range contingencies between events. Subjects were exposed to sequences that contained event-to-event dependencies such that the actual event appearing on any Trial n was stochastically dependent upon the event that had occurred on some previous Trial $n - j$ where $j = 1, 3, 5$, or 7. Training consisted of several hundred trials of the instant asymptote procedure with the particular stochastic dependency for that session. During the first session $j = 1$, during the second $j = 3$, and so forth.

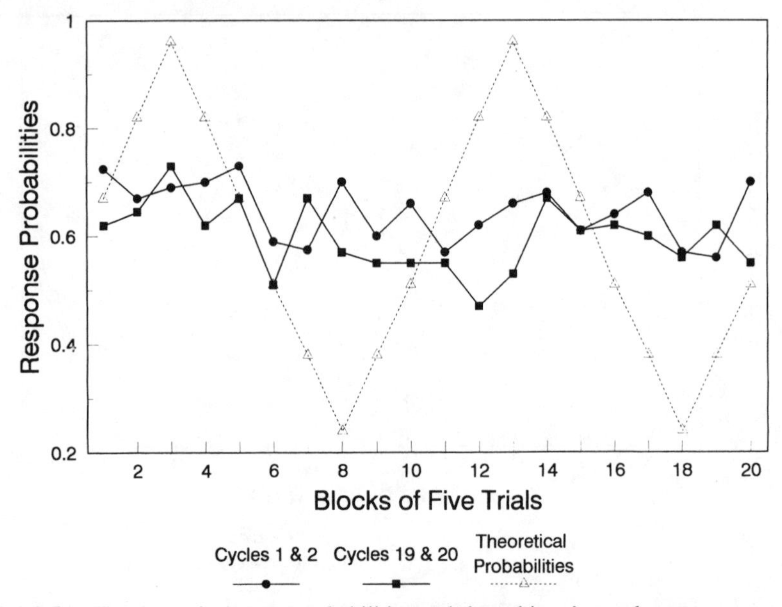

Fig. 2.5. The theoretical event probabilities and the subjects' actual response proportions in blocks of five trials for the first two cycles and the 19th and 20th cycles of testing on Day 2 of the experiment. This figure is adapted from "Event tracking in probability learning" by A. S. Reber and R. B. Millward, 1971, *American Journal of Psychology, 84,* p. 89. Copyright 1971 by the American Journal of Psychology.

During testing, subjects displayed a clear sensitivity to these dependencies, a sensitivity reflecting an ability to exploit structure that required knowledge of event dependencies as remote as seven trials. What makes this finding impressive is that this capacity appears to be beyond what were found in earlier work (Millward & Reber, 1968; Reber & Millward, 1965) to be limits on explicit recall. In those earlier experiments, subjects were asked to recall which event had occurred on a specified previous trial. Beyond $n = 5$, they were virtually reduced to guessing. Hence, we have one of the classic elements of implicit learning: *Implicitly acquired knowledge is responsible for performance that goes beyond, as it were, what estimates of conscious knowledge would predict.*

CONTROL-TASK STUDIES. Similar observations concerning the exploitation of structure have been made in somewhat different contexts by other researchers. Broadbent and his colleagues have shown that knowledge of complex rule systems governing simulated production and social systems is also acquired and utilized in an implicit fashion (Berry & Broadbent, 1984, 1987, 1988; Broadbent, 1977; Broadbent & Aston, 1978; Broadbent et al., 1986; Hayes & Broadbent, 1988). In these studies subjects are presented with either an imaginary manufacturing situation, such as a sugar production plant or a "computer person" who displays various social traits. In the production-control experiments, a subject may be told that he or she is the manager of a sugar production plant

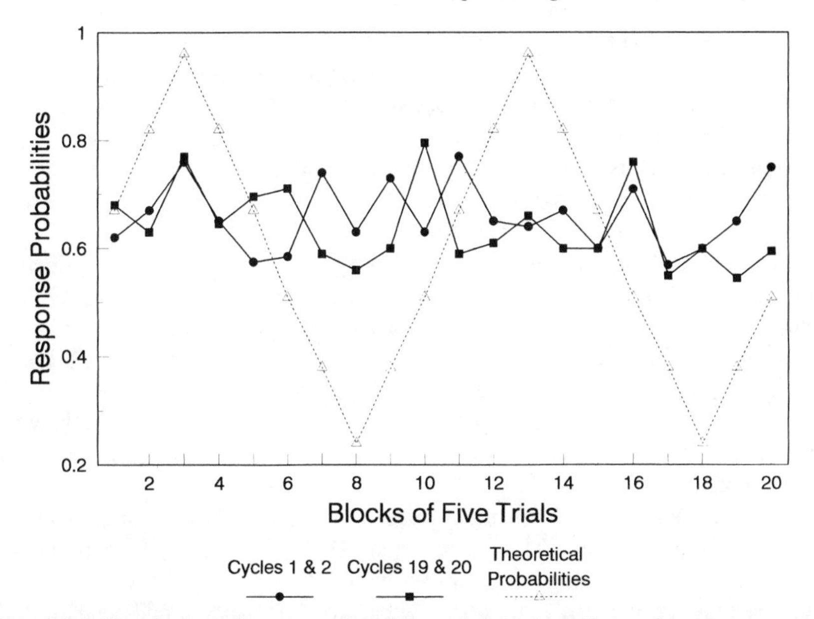

Fig. 2.6. The theoretical event probabilities and the subjects' actual response proportions in blocks of five trials for the first two cycles and the 19th and 20th cycles of testing on Day 3 of the experiment. This figure is adapted from "Event tracking in probability learning" by A. S. Reber and R. B. Millward, 1971, *American Journal of Psychology, 84,* p. 89. Copyright 1971 by the American Journal of Psychology.

and asked to modify production to achieve some preassigned level. The control is implemented by making adjustments in one (or, on occasion, more) elements of production, such as wages, labor peace, number of workers hired, and the like. The experiments are run on a computer, which is programmed with the rules that relate production levels to these other factors.

The subject's task is to make shifts in the value of the relevant parameter in an attempt to achieve the target production level. The computer interacts with the subject by displaying production levels in response to parametric adjustments. A typical rule is something like: $A = 2w - (p + n)$, where A is the output value, w is the number of workers employed, p the previous trial's output, and n a noise factor designed to introduce a measure of indeterminacy in the task. Note that because of the noise factor, perfectly accurate responding cannot be achieved even if the subject consistently uses, implicitly or explicitly, the proper rule. Hence, in these studies, responses that are ± 1 level with respect to the target are regarded as correct. Ceilings are set, typically representing the lowest and highest of 12 levels, and output production cannot go beyond these. Many other rules of varying complexity, of course, may easily be implemented.[3]

[3] For example, in the Broadbent and Aston (1978) paper in which the subjects were students in a senior management course, the production rule was extremely complex. The subjects worked in teams and attempted to maximize the overall economic welfare of a country using a multi-

In the social-control experiments, the task is set up so that it is an exact analog of the production-control task. However, rather than manipulating sugar production, the subject interacts with a computer person (herein known as "Max"), who displays any number of emotions ranging, for example, from rude to loving. The subject's task is to get Max to respond in a particular manner by manipulating his or her own affective reactions to Max. Max's mood is modified according to the same rule as in the production task.

These control tasks are intriguing. They are more "real worldly" than either the grammar-learning or the probability-learning procedures. Yet, they have most of the properties outlined above for an experimental setting within which to explore implicit learning. In addition to the work of Broadbent and his research group, they have been employed by Mathews, Stanley, and their colleagues in a series of experiments designed to explore the relationship between usable knowledge and verbalizable knowledge, which will be discussed later in this chapter.

In these studies, data that parallel those from the grammar-learning and probability-learning experiments were reported. Achieving the required production standards or manipulating successfully the affective tone of Max requires that the rules be "known" in some sense of that word. Broadbent and Mathews and their colleagues consistently report that subjects make appropriate adjustments to the parameters indicating that they are capable of exploiting the rules and that they do so primarily in the absence of conscious knowledge of the rules themselves. Indeed, in several experiments where the rules were complex and not particularly salient, the capacity to articulate rules was found to be negatively associated with actual performance (Berry & Broadbent, 1987). This relationship between the capacity to express a rule verbally compared with the capacity to use it appropriately will occupy us at some length.

There is an interesting footnote to this research program. The social tasks in which subjects have to modify Max's mood typically produced better performance than the manufacturing tasks. This difference is of some interest in that it suggests that pragmatic considerations may be as important in implicit situations as they are in explicit. To my knowledge, there has not been a systematic exploration of this factor.

MATRIX-SCANNING STUDIES. Additional data that support the general picture of implicit learning were reported by Lewicki and his co-workers (Lewicki et al., 1987; Lewicki et al., 1988). In the Lewicki et al. (1987), for example, study subjects were confronted with a complex matrix of numbers arranged in four

---

parameter formula, which had been developed to provide a simulation of the economy of the UK. It included such parameters as imports, exports, inflation, unemployment, the GNP, marginal tax rates, government expenditures, and the money supply. In the usual study in cognitive psychology, rules like this are not employed although there is, in principle, no reason why they should not be. There are many reasons to think that it would be profitable to look in depth at such complex learning processes within a laboratory setting.

quadrants on a computer screen. The matrix contained a single target number whose location changed from trial to trial. The subjects' task was to note the location of the target number by pressing, as quickly as possible, a button indicating the proper quadrant.

The key trial was the last of each run of seven. On that trial the location of the target was determined by the sequence of locations that it had appeared on the first, third, fourth, and sixth trials of that run (the locations on the second and fifth trials were selected at random). For example, if the target appeared in Quadrants 3, 1, 4, 2 on the critical trials, it would always occur in Quadrant 1 on the seventh trial; if it appeared in Quadrants 4, 1, 3, 2, then it would appear in Quadrant 3, and so forth. The relationships between the sequence and the test trial were arbitrary but consistent; the location of the target on Trial 7 was always determined by its locations on Trials 1, 3, 4, and 6. Over several thousand trials (like some of our PL studies already described, learning relationships like these can take considerable practice) subjects' reaction times (RTs) and error rates on the critical seventh trial decreased significantly. To ensure that the subjects were not simply acquiring a sensorimotor task, after the RTs stabilized (around 4000 trials) the contingencies were changed—producing an abrupt increase in RTs and error rates.

This learning appeared to be completely outside of consciousness. Subjects were unaware of the existence of the both the original pattern and the rule shift. When the subjects were asked to predict overtly where the target would appear on the seventh trial, they were no better than chance. Moreover, when asked, no subject was able to identify which of the six trials were the critical ones or to give a coherent characterization of the principles involved in determining target location.

In the Lewicki et al. (1988) a similar procedure was used in which continuous sequence of targets appeared in one of the four quadrant. Here, however, the sequence was constructed according to "logical" blocks of five trials each. Within each set of five trials, the first two target locations were determined pseudorandomly (the only restriction being that the targets could not both occur in the same location); the final three target locations were dictated by a biconditional rule such that each location was determined by the locations of the two preceding targets. As with the earlier study, subjects gradually responded with shorter and shorter RTs to the rule-governed events (the third, fourth, and fifth of each logical block) but showed no such improvement with the pseudorandom events (the first and second). An unannounced shift in the rule governing event locations had the expected effect of slowing RTs to the rule-governed targets but having no effect on the pseudorandom ones.

Here, too, the learning was clearly implicit in that no subject was able to articulate the rules that they appeared to be using—even when they were offered a $100 reward for any coherent characterization of the patterns that dictated the sequence of target locations. As an interesting little touch, in this study all of the subjects were faculty colleagues in Lewicki's psychology department and all knew that the research was oriented toward the study of nonconscious cognitive processes.[4]

These matrix-scanning experiments, along with a series of studies by Nissen and her colleagues (Knopman & Nissen, 1987; Nissen & Bullemer, 1987; Nissen et al., 1987) in which subjects work with a repeating sequence of lights, are generally taken as supporting the overall notion that subjects come to exploit the structure of the environment. In Nissen's experiments,[5] subjects work with a simple visual stimulus (e.g., an asterisk) that flashes in one of four locations on a screen. The location shifts among the four in a sequence of length 10, which repeats ten times for a total of 100 consecutive flashes. The subjects' task is roughly the same as it is in the matrix-scanning studies; they must press the button that corresponds to each stimulus event as quickly as possible. The typical finding here is similar to that in Lewicki's studies; subjects get quite good at the task independent of conscious knowledge of the actual sequence in use.

We should note that there is a methodological difference between these matrix-scanning and sequence-scanning studies and the grammar-learning, probability-learning, and production-control tasks. The stimulus sequences in Lewicki's and Nissen's experiments are fixed. They are complex to be sure, but they are fixed. That is, in the Lewicki matrix-scanning experiments if the critical stimulus appears in say, Quadrants 1, 2, 4, and 3 on Trials 1, 3, 4, and 6, then it will definitely appear in Quadrant 1 on Trial 7; in the Nissen sequence-learning studies, the sequence of 10 locations is selected accordingly to particular constraints, but once established it is fixed and is presented repeatedly. The "fixing" is greater in the case of Nissen's procedure than it is in Lewicki's, but neither of them are using stimuli whose underlying nature is abstract.

Hence, these studies should probably be seen in a different light from the others. They clearly support the overall argument of this section that people come to learn to use the structure inherent in the stimulus environment to guide

[4] The findings of the Lewicki, Hill, and Bizot (1988) study were challenged in a paper by Perruchet, Gallego, and Savy (1990). Perruchet and his colleagues pointed out that there was an artifact in the original design such that the stimulus sequences were not necessarily to be resolved as logical blocks of size five and, hence, that the rule that Lewicki et al. had used to construct them was not necessarily the rule that subjects had learned. The artifact emerged from the recognition that the Lewicki rule produces a sequence of "movements" of target from location to location that is not uniform; some movement patterns are much more likely than others, especially on the final three trials of each "logical" block. Perruchet et al. replicated Lewicki et al.'s basic findings and, by examining the fine grain of the data, were able to show fairly convincingly that subjects were not learning the Lewicki rule but rather were responding to patterns of frequency of movement of the targets.

However, the discovery of this artifact in no way diminishes the impact of these studies; subjects may not be learning the rule that Lewicki et al. thought they were, but they were certainly learning to respond to a complexly structured stimulus array and were doing so independent of awareness of what that structure was. This issue is of some importance and is pursued in more detail in the section "On Rules" in Chapter 4.

[5] The work of Nissen and her colleagues has also been used to examine the differences between implicit and explicit processes in various special populations including those with neurological and psychiatric disorders. This element of her research program will be discussed in more detail in Chapter 3 where the question of the robustness of the implicit cognitive system is explored in the context of evolutionary considerations.

their behavior, although the learning here is rather more "concrete" than that which occurred in the grammar-learning or the probability-learning experiments in that in these latter studies' subjects are acquiring abstract knowledge of rule-governed systems.

In this context, a recent study by Cleeremans and McClelland (1991) is interesting. They blended the procedures of Nissen and Lewicki with that of the traditional artificial grammar learning in a novel fashion. Subjects were run extensively using a highly complex event sequence. Each event was a light that flashed at one of six locations on a monitor screen. The subjects were asked to respond to each event by pressing a button that corresponded to the location of the flash as quickly and accurately as they could. However, the stimulus sequences were neither arbitrary nor fixed as in Lewicki's and Nissen's studies; they were generated by the transition rules of a "noisy" finite-state grammar. Actually, they were even more complex than the stimuli used in traditional grammar-learning studies in that 15% of the events were random and did not follow the rules of the grammar (hence, the "noisy" grammar). After more than several thousand trials, subjects showed clear learning of the underlying structure of the display as evidenced by dramatic decreases in RTs to grammatical events compared with RTs to the events generated at random.

SUMMARY. Clearly, subjects learn to utilize the structural relationships inherent in these various complex stimulus domains. No real surprises here. In many ways these various studies serve as complicated existence demonstrations. They show that it is possible to obtain this kind of unconscious, nonreflective, implicit learning in a controlled laboratory setting, that it can occur in a relatively short time span, and that it can be seen to emerge when the stimulus is a structured domain whose content is arbitrary and distinctly remote from typical day-to-day experiences with the real world.

On implicit versus explicit processes

The experiments reviewed to this point were all run under instructional sets in which subjects were not informed that the stimuli were structured or rule defined. In these cases, the point was to maximize the emergence of implicit learning. The (implicit) assumption was that the implicit induction of knowledge is the default mode—all things being equal.

It is important to be clear about this issue and the kinds of manipulations that have been used. It is universally accepted that the college undergraduate whose cognitive processes form (for better or worse) the foundations of our science is an active and consciously probing organism, especially when it comes to things with structure and patterns. The above experiments were carried out by circumventing this pattern-searching tendency in any number of ways. Reber and Millward's PL studies, Nissen's sequence-learning studies, Lewicki's target location experiments, and Cleeremans and McClelland's event sequence procedure all accomplished this by "blitzing" the subjects with information at rates beyond those at which conscious code-breaking strategies could operate. The artificial grammar-learning studies and Berry and Broadbent's production-control and social-control system experiments were successful because the structure of the stimuli was highly complex and the instructions to the subjects were calculatingly vague. An obvious question is what effect would explicit instructions have? What happens when subjects are informed, at the outset, that the materials they will be working with reflect regularities and patterns?

The first manipulation of the factor of explicitness used the PL technique (Reber, 1967b; Reber & Millward, 1968). The procedure consisted simply of telling some of the subjects exactly what was going on in the experiment. Specifically, one group of subjects was given concrete instructions about the frequency characteristics of the event sequence they would be asked to predict by informing them of the relative probabilities of the two events. These subjects were then run on a standard PL procedure and compared with a control group run using identical event sequences but without the explicit information.

The information about the event frequencies had virtually no effect on behavior as can be seen in Figure 2.7. The two groups were statistically indistinguishable from each other, even on the first block of 25 prediction trials where the impact of the instructions would have been most likely to be felt. Clearly, prob-

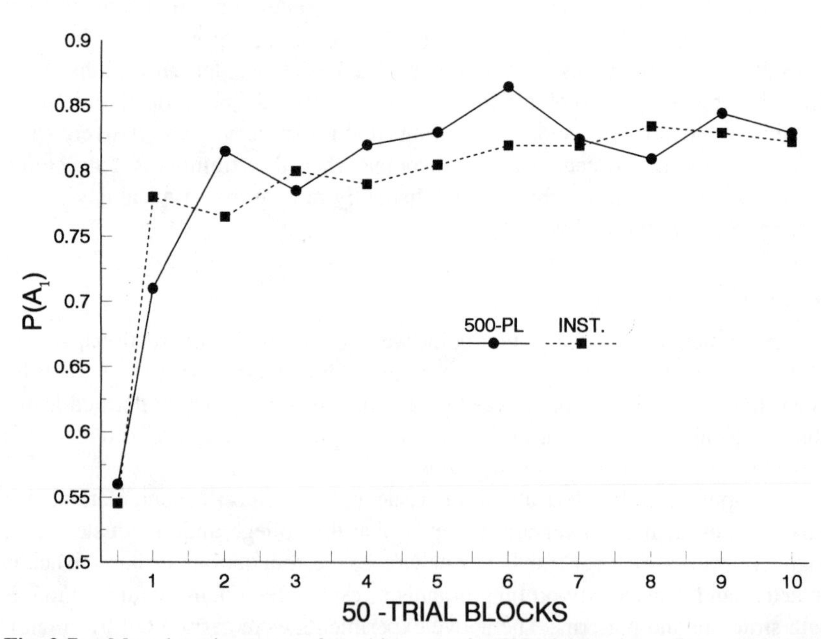

Fig. 2.7. Mean learning curves for subjects run for 500 trials under ordinary probability learning conditions (500-PL) and for subjects first given detailed instructions concerning the probabilistic nature of the task (INST). This figure is adapted from "Event observation in probability learning" by A. S. Reber and R. B. Millward, 1968, *Journal of Experimental Psychology, 77,* p. 321. Copyright 1968 by the American Psychological Association.

ability learning is more than the learning of probabilities. Instead, what is really going on is the apprehension of deep information about the nature of the structure of the sequence of events.

Postexperimental debriefings were revealing. Subjects were quite clear about knowing which light would be the dominant one, and all said they believed the instructions. But, in virtually every case, they claimed that somehow the specific information lacked meaning that they felt they could use. It took real experience with the event sequence to impart a knowledge base that was usable for directing choices on individual trials. Note that this experiment used Bernoulli sequences; there was no "structure" in the usual sense of the word. Nevertheless, subjects reported achieving a sense of the nature of the event sequence from experience with events that they did not derive from the explicit instructions. Importantly, this occurred despite the fact that, in principle, there is nothing to be extracted from the event sequence other than the relative frequencies of the two events.

Several studies using the grammar-learning procedure have been run that also explore the boundary between implicit and explicit knowledge. The first of these used the simple device of encouraging one group of subjects to search for the structure in the stimuli while a comparable group was run under a neutral instructional set (Reber, 1976). Both groups were given the same learning phase, during which they had to memorize exemplars from a synthetic grammar and an identical testing phase during which they were asked to assess the well-formedness of novel letter strings. Unlike the earlier procedure, in this study informed subjects were told only about the existence of structure, nothing was said about the nature of that structure.

The well-formedness test used in this experiment is one of the standard procedures we developed to explore the knowledge base that subjects take from the learning experience. The concept of well-formedness is used here exactly as it is in logic or linguistics. Given any formal grammar, a symbol string is, by definition, well-formed if that grammar could generate the string. That is, well-formed strings are grammatical strings. By asking subjects to judge the well-formedness of novel strings we can get a pretty good sense of the knowledge they have of the rules of the grammar.

The explicitly instructed subjects in this study performed more poorly in all aspects of the experiment than those given the neutral instructions. They took longer to memorize the exemplars, they were poorer at determining well-formedness of test strings, and they showed evidence of having induced rules that were not representative of the grammar in use. The suggestion is that, at least under these circumstances, implicit processing of complex materials has an advantage over explicit processing.

However, as gradually became clear, what this study actually showed is that explicit processing of complex materials has a decided disadvantage relative to implicit processing. This is no mere play on words. The implicit/explicit distinction is rather more complex than it first appeared. Analysis of the fine grain of the data from the Reber (1976) article revealed that the explicit instructions seemed to be having a particular kind of interference effect. Specifically, subjects were being encouraged to search for rules that, given the nature of finite-state grammars with their path-independent, Markovian properties and given the kinds of attack strategies the typical undergraduate possesses, they were unlikely to find. Moreover, they tended to make improper inductions, which led them to hold rules about the stimuli that were, in fact, wrong. The simplest conclusion here seems to be the right one: *looking for rules won't work if you cannot find them*.

In a number of other studies, instructions of various kinds have been shown to have any of a number of effects. Berry and Broadbent (1988), using the control-task procedure, reported that explicit instructions were counterproductive under the usual experimental setting, but when the rules in use were made highly salient, the instructions were beneficial. Brooks (1978) used finite-state grammars similar to the one used in Reber (1976) and a paired-associates learning procedure in which strings of letters from grammars were paired with responses of particular kinds (e.g., animal names, cities). He found that informing subjects about the existence of regularities in the letter strings lowered overall performance. Reber, Kassin, Lewis, and Cantor (1980) found poorer performance with explicit instructions when the stimuli were presented in a large, simultaneous array in which letter strings were posted on a board in haphazard fashion. Howard and Ballas (1980) reported detrimental effects of explicit instructions using structured sequences of auditory stimuli when there was no systematically interpretable pattern expressed by the stimulus sequences. In all these cases, the original finding was basically replicated.

However, as a variety of different studies are examined the picture gets murkier instead of clearer. Briefly: Millward (1981) found no difference between explicitly and implicitly instructed subjects in an experiment that, in principle, looked like a replication of Reber's 1976 paper with the seemingly modest variation that the strings used during learning were up to eleven letters long, whereas the stimuli in Reber's study were no longer than eight letters. Cantor (1980) found inconsistent effects of instructions when number strings were used instead of letter strings even though they were generated by the same Markovian grammar. Abrams (1987) and Rathus, Reber, and Kushner (1990), using what were replications of Reber (1976), failed to find the instructional effect. Dulany et al. (1984), working with synthetic languages, reported no significant differences between the two instructional sets although in this case a rather different testing procedure may have masked differences. Mathews, Buss, Stanley, Blanchard-Fields, Cho, & Druhan, (1989) found a complex pattern of differences between instructional groups depending on whether the letter set used to instantiate the grammar was modified over the several days of the experiment. Their grammars were also somewhat more complex than what most others have used. Danks and Gans (1975) reported no differences using a synthetic system that was considerably simpler than the Markovian systems used by others.

Finally, there have been several studies that showed an advantage for the explicitly instructed subjects. Howard and Ballas (1980) reported that the explicit instructions that debilitated performance when introduced under conditions of semantic uninterpretability could also function to facilitate performance when the stimuli expressed semantically interpretable patterns. Reber et al. (1980) showed that it was possible to shift performance rather dramatically by intermixing the instructional set with the manner of presentation of the stimulus materials and with the point at which explicit instructions were introduced during learning. The fact that several of these studies were done in our own laboratory (including the Cantor, the Abrams, and the Rathus et al. nonreplications) adds, shall we say, a bit of spice to the situation.

However, there are two factors here that help make these data somewhat less haphazard than they appear to be. The first is psychological salience; the second is the circumstances under which the instructions are given to the subjects. The first of these is the more interesting and the one from which we can gain insight into process.

In several of the instances where explicit instructions facilitated performance, the manner of presentation of the stimuli was such that the underlying factors that represent the grammar were rendered salient. In Howard and Ballas (1980), the semantic component focused the subjects on the relevant aspects of the patterned stimuli. The effectiveness of such a semantic component has been often noted in artificial-grammar learning studies (Moeser & Bregman, 1972; Morgan & Newport, 1981). In Reber et al. (1980), the simple expedient of arranging the exemplars of the grammar according to their underlying form so as to increase the salience of the rules produced the instructional facilitation. It should be noted here that the Berry and Broadbent experience with manipulating salience of the underlying rules in a very different setting lends additional support to this line of research.

Moreover, several other studies seem to have arranged matters inadvertently so that structural properties were made more salient. For example, in Millward (1981) the use of longer strings provided many opportunities for subjects to be exposed to the loops or recursions in the grammar (see Figure 2.1) and thereby increased the psychological salience of the underlying structure. In the Cantor (1980) experiment subjects reported treating the number strings as puzzles akin to the sequence problems found in supplements to the Sunday newspaper. Such a cognitive set invites explicit hypothesis testing from subjects rather than implicit induction processes. In Danks and Gans (1975), the relatively simple nature of the stimuli likely acted to equate the mode of processing of the stimuli in both groups. That is, both were likely using a reasonably explicit mode independent of the instructions. Thus, the converse of the earlier conclusion: *Looking for rules will work if you can find them.*

There are cases that appear to be genuine failures to replicate the original finding, specifically Dulany et al. (1984), Abrams (1987), and Rathus et al. (1990). In Dulany et al. the procedure used during learning seemingly should have yielded a difference during testing. There are some interesting aspects of this experiment, but there are no obvious reasons why the effect failed to emerge. On the surface, the Abrams and the Rathus et al. studies should also have produced an instructional effect.

In any event, the literature on the implicit/explicit problem is clearly complex, and it takes but a moment's reflection to appreciate that there are still other issues lurking behind these findings. First, it seems clear that any number of confounding factors may influence, either positively or negatively, the impact of explicit instructions. Such instructions may introduce an element of stress or anxiety, they may evoke a sense of motivation to succeed on the task, encourage one or another conscious strategy, and the like. For example, Kassin and Reber (1979) found that performance on a grammar-learning task was systematically related to the locus of control dimension. Subjects who were high on the "internal" scale tended to be better learners. However, to date very little of the work here has taken such factors into account, nor is much is known about the manner in which conscious, explicit, processing systems interact with the implicit and unconscious.

Second, it seems clear that we are still dealing with a somewhat limited kind of analysis of complex learning, particularly if one wishes to view this research in its constrained, laboratory setting as representing a kind of general metaphor for real-world acquisition processes. As I have argued, there is no reason for assuming that implicit and explicit systems are functionally distinct. It is most assuredly the case that they are richly interactional, and under most real-world settings complex skills are acquired with a blend of the explicit and the implicit, a balance between the conscious/overt and the unconscious/covert. In the end, the instructional variable may turn out to be a weak one and we should not be surprised if the simple manipulations used produce occasionally ambiguous results.

Moreover, there is surely a difference between simply informing a learner that the stimulus materials have structure and telling the learner something definitive about that structure. The next section deals with this issue.

The effects of providing explicit information

The issue here concerns the impact of giving subjects precise information about the nature of the stimulus display that they will be exposed to. This is an important question for a number of reasons. For one, this issue broaches on some of the classic questions in pedagogic theory about how best to communicate to students highly complex and richly structured information. It also emerges in various studies on the acquisition of expertise in such areas as medical diagnosis where the relationship between specific knowledge presented to medical students and their emergent tacit knowledge base is turning out to be most complex (see, e.g., Carmody, Kundel, & Toto, 1984). Eventually this issue will have an impact on the expert/novice problem. Or to phrase it another way, eventually those working on the expert/novice problem are going to have to come to grips with the work on implicit learning. One of the most difficult problems facing workers in expert systems is discovering what it is that the human expert knows. In the early days of this field it was naively assumed that one merely had to ask the expert. This, unfortunately, turned out to be a futile gesture when knowledge is held implicitly.

In Reber et al. (1980) an attempt was made to address the general issue directly using the standard grammar-learning procedure. In that study, subjects were presented with the actual schematic structure of the grammar; that is, they were presented with Figure 2.1. Each subject was handed a copy of the diagram and given a seven-minute "course" in how such a structure can be used to generate strings of symbols. This procedure was supported by an observation period during which a set of 60 exemplars (20 distinct strings each displayed three times) was shown to the subjects. This training format thus mixed a maximally explicit learning procedure with a maximally implicit one.

The manner of interaction between these two modes of apprehension was explored by introducing the explicit training at different points in the observation period. One group of subjects received the explicit instruction at the outset before any exemplars were seen; one group received it partway through the observation period; and a third group had the explication of structure delayed until after the full set of exemplars had been observed. As in the typical grammar-learning study, knowledge acquired during learning was assessed by a well-formedness task.

The key finding was that the earlier during the observation training the explicit instructions were given, the more effective they were. From the previous discussions it is clear that increasing the salience of the relationships between symbols increases the effectiveness of subjects' attentional focus. It is also clear that instructions that encourage the subjects to deal with the stimuli in ways that are discoordinate with underlying structure have detrimental effects on acquisition. Thus, the explication of the precise nature of the structure underlying the stimuli must have differential impact on learning depending on which of these two processes is encouraged.

The most plausible interpretation here, and the one that has interesting implications for theories of instruction, is that the function of providing explicit instructions at the outset is to direct and focus the subjects' attention. It alerts them to the kinds of structural relations that characterize the stimuli that follow and permits appropriate coding schemes to be implemented. Yet, these instructions did not teach the grammar in any full or explicit fashion; instead they oriented the subjects toward the relevant invariances in the display that followed so that they, in effect, taught themselves.

Accordingly, when such explicit instruction is introduced later in the observation period, its effects are different because two sources of difficulty are introduced. First, it imposes a formalization of structure that is, in all likelihood, discoordinate with the tacit system that was in the process of being induced. Second, it reduces the number of exemplars that can be used as a base for extracting invariance patterns. In the extreme case where the instructions were delayed until the completion of the observation period, this informational base is virtually eliminated.

Several of the researchers who have worked with the Broadbent control tasks have looked at this relationship between performance and explicit instruction. Berry and Broadbent (1984) found that giving subjects very explicit instructions about how to control sugar production and/or the personal interactive style of Max produced no increase in their performance with either task. Interestingly, it did improve their scores on the postexperiment questionnaire, thereby reducing even further the relationship between explicit knowledge and implicit knowledge. This result parallels the Reber and Millward (1968) finding with the PL procedure discussed above in that the subjects clearly believed and understood the instructions but found them to be of little or no value in directing their behavior—experience with the materials is what counted.

Stanley et al. (1989) provided additional evidence for this position. They basically replicated and extended the Berry and Broadbent findings by giving subjects a variety of different kinds of explicit information about the tasks. These instructions ranged from providing subjects with a precise heuristic that, if followed, guaranteed success in the task to giving subjects a statement taken from another subject who had been through the learning procedure and had scored highly on the task.[6] Stanley et al. reported some moderate effects of the instructions in that the instructed subjects were above chance on a follow-up test. However, no differences between the various forms of instruction were found, and all groups of instructed subjects performed far below the subjects who had been given a standard learning experience. As Stanley et al. concluded, "the dissociation between verbal knowledge and task performance appears to be less than total, but much deeper than found in most other cognitively demanding tasks."

From the preceding discussion, it should be clear that there are fairly deep issues lurking here, and they can use some further exploration. For one, there is every reason to suspect that subjects' tacit representation of rules is idiosyncratic in various characteristics—and that this idiosyncratic quality is going to be independent of the task. From earlier work, it is clear that subjects are known to use a wide variety of coding schemes in focusing their attention on the stimuli. In studies where there have been extensive postexperiment debriefings (Allen & Reber, 1980; Reber & Lewis, 1977) or where subjects provide an ongoing introspective analysis (Reber & Allen, 1978), we found evidence of rich and often wonderfully idiosyncratic "devices" introduced on the part of our subjects. They ranged from mnemonic devices based on hagiographic codes (from a subject who was preparing to enter divinity school) to coding systems linked with the initials of girlfriends.

At first we were startled by the extreme diversity of these mnemonics. How could such bizarrely disparate coding schemes yield such a common data base? The answer turns out to be, in fact, rather simple. The schemes are, in large measure, irrelevant to the final mental representations. So long as they do not entail inappropriate rule formation, their impact is superficial. Independent of individualistic mnemonics, attentional focusing priorities, or preferred rehearsal

[6] This technique of using one subject's explicit statement to assist a naive subject represents an intriguing version of the classic yoked-control procedure. It is used effectively in Stanley, Mathews, Buss, and Kotler-Cope (1989) to explore the implicit/explicit relationship in these control tasks and in Mathews, Buss, Stanley, Blanchard-Fields, Cho, and Druhan (1989) to examine similar factors in the artificial-grammar learning task. It is discussed in more detail later.

strategies, the implicit learner will emerge from the training session with a tacit, valid knowledge base coordinate with the structure of the stimulus environment.

The degree to which the explicit instructions introduce difficulties will thus be dependent on the extent to which the subject's tacit representation of structure matches the formalization provided by the schematic of the grammar and the accompanying characterization of its generational properties. In Dulany et al.'s (1984) terms, subjects are learning "correlated grammars" whose properties are, in all likelihood, not commensurate in any simple way with the Markovian system in use. Recent results from Mathews et al. (1988) strongly corroborate this interpretation.

The reasons for the relatively poor performance of the group that received the formal training at the end after the full observation period in Reber et al. (1980) should now be apparent. The difficulty is that there is a potentially infinite number of formalizations that could account for the structure displayed in any given subset of strings from one of these grammars; present the "wrong" one to a subject and the instructions will not have a salutary effect. How much can we expect a subject to benefit from the specific information that the set of exemplars just observed and tacitly coded as, say, bigram covariation patterns based on the names of saints is "in reality" to be formalized as a Markovian process? To take an obvious analogy, most of us with our extensive observation and generation of utterances in English have failed to derive any facilitative effect of explicit instruction with transformational grammar, which at least in principle, can be posed as a legitimate formalization of our tacit knowledge. Moreover, such explicit awareness of structure can actually be a nuisance when one tries to fulfill the kinds of demands placed on subjects in these experiments as in the discrimination of well-formed, novel instances from those that contain some violation of the formal system.

Berry and Broadbent (1988) reported data that give additional force to this line of argument. Using their standard control procedure, they transferred subjects from one task to another to see if knowledge acquired in one could be used in the other. Both tasks used the same underlying rule, although the manner in which it was instantiated differed, with one personal and the other impersonal. When subjects were shifted between similar tasks (i.e., between two personal or two impersonal tasks) and operated under an implicit instructional set, transfer was observed. However, subjects showed no such transfer when they were previously informed of the critical relationship between the tasks. Berry and Broadbent argued that, just as in studies with artificial grammars, subjects' implicit knowledge base is used to direct behavior and control effective action, independent of conscious representations. It should also be noted, however, that unlike other studies (see the section on Knowledge Representation in Chapter 4) Berry and Broadbent failed to find transfer across dissimilar tasks; subjects did not transfer when the shift was between personal and impersonal tasks.

In summary, although there are not a lot of hard empirical data here, those that are available point toward an interesting conclusion. Specific instruction concerning the materials to be learned in complex situations will be maximally beneficial when there is representational coordination between it and the tacit knowledge derived from experience. I take this whole line of research to support the general line of argument that there is no necessary connection between the underlying epistemic form of the knowledge that our subjects acquire in an implicit learning task and our formalizations of the task. Because this issue will ultimately be of importance for theories of instruction, it is one that is in need of closer examination than it has, to date, received in the field.

On deep and surface structures

The issue here is the degree to which implicit learning can be seen as acquisition of knowledge that is based on the superficial physical form of the stimuli or as knowledge of the deeper, more abstract relations that can, in principle, be said to underlie them. I want to deal with this separately from the more complex and important problems of mental representation, which are discussed in the next section, because the issue has been addressed directly in the literature and the manner in which it is resolved has clear entailments for the representation issue. I hope that this will all become clear in a paragraph or two.

In an early paper (Reber, 1969) evidence was reported for the proposition that implicit knowledge is abstract and not dependent in any important way on the particular physical manifestations of the stimuli. This study consisted of two sessions during which subjects memorized letter strings from a grammar. When the second session began the stimulus materials were, without warning, modified. For some subjects the same letters continued to be used but the rules for letter order were now those of a different grammar (the syntax, if you will, was altered). For other subjects the underlying structure was not tampered with but the letters used to represent the grammar were replaced with a new set (the vocabulary was changed). The two obvious control groups, one where both aspects were altered and one where neither was changed, were also run. As can be seen in Figure 2.8, the various manipulations had systematic effects upon subjects' ability to memorize stimuli in the second session. Modification of the rules for letter order produced decrements in performance; modifications of the physical form had little or no adverse effects. So long as the deep rules that characterized the stimuli were left intact, their instantiations in the form of one or another set of letters was a factor of relatively little importance.

A study by Mathews, Buss, Stanley, Blanchard-Fields, Cho, and Druhan (1989) supported this general finding. Their experiment was an extensive synthetic-grammar learning study run over a four-week period and contained a number of complex elements, one of which was a shifting of the letter set used to instantiate the grammar. Subjects who received a new letter set each week of the experiment (based on the same underlying syntax) performed as well as subjects who worked with the same letter set throughout the course of the experiment. The effect here was quite striking; the transfer from letter set to letter set occurred, using their words, "immediately and automatically without any conscious translation process."

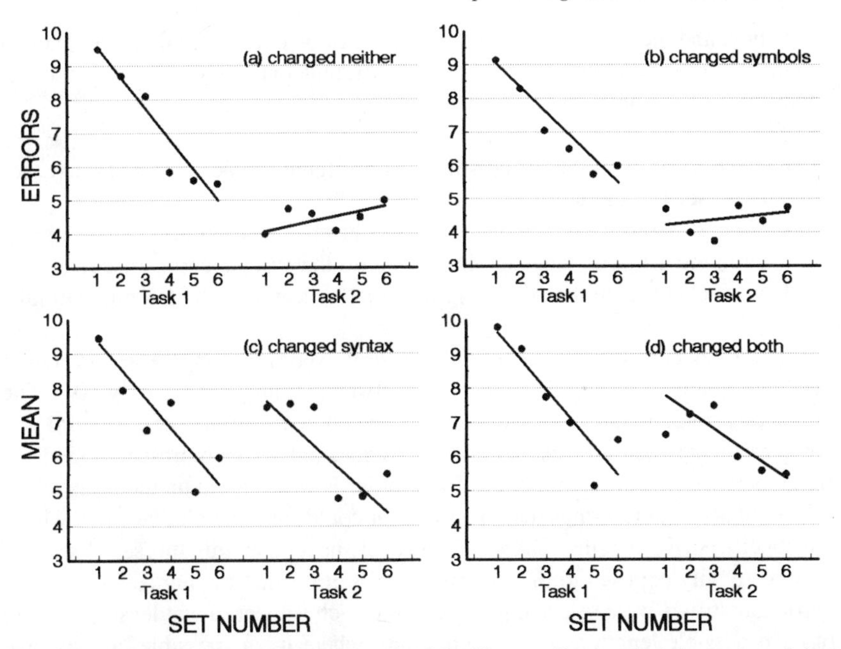

Fig. 2.8. Mean errors to criterion on each of the 12 sets of items. Task 1 was the original learning condition, Task 2 was after the break when the changes were introduced. This figure is adapted from "Transfer of syntactic structure in synthetic languages" by A. S. Reber, 1969, *Journal of Experimental Psychology, 81,* p. 117. Copyright 1969 by the American Psychological Association.

Reber and Lewis (1977) reported an equally striking example of the abstract nature of tacit knowledge. In that study, knowledge of the grammar was assessed by having subjects solve anagram problems. After a standard training session during which subjects memorized exemplars from the language, they solved anagrams from the synthetic language over a four-day period. This task consisted of giving subjects a stack of tiles with letters on them and asking them to arrange the letters to make an acceptable string in the language. They were told that the items they memorized during the learning phase could be taken as examples of what acceptable strings looked like. There was no feedback about the correctness of their solutions.

This procedure produces, as one might guess, a great deal of data. For reasons discussed below, it is convenient to code the letter strings that subjects generated as their solutions in the form of bigrams and to note the rank order of frequency of occurrence of each possible bigram. For example, suppose a subject produced the string PTTTVV as a possible acceptable string. This string contains bigrams PT, TV, and VV once each and TT twice. In this manner every string can be represented as a set of these bigrams. The total set of proffered solutions, as well as the actual strings of the underlying grammar, can be coded in terms of bigram frequency, and the rank order of the bigrams can be obtained. Given the manner in which this experiment was run, three different rank orders of frequency of occurrence of bigrams exist:

Rank order A based on the actual solutions proffered by the subjects (corrected, of course, for guessing).
Rank order B based on the frequency of occurrence of each acceptable bigram within the artificial language itself (within the strings lengths used).
Rank order C based on the actual bigrams that appeared in the learning stimuli.

Rank order correlations between these three were revealing. The correlation between A and B was .72, while that between A and C was only .04. The interesting point about these results is that the comparison between A and B is a comparison between subjects' usable knowledge and a deep representation of the frequency patterns of the grammar. Rank order B was formed on the basis of the full set of acceptable strings that the grammar could, in principle, generate within the specific string lengths. Subjects, however, never saw this full set of strings; they were only exposed to the exemplars chosen for the training session. These particular strings were selected to ensure that each subject saw at least one string of each possible length and, for each length where it was possible, at least one example of each of the grammar's three loops. This procedure yielded a set of strings that displayed the deep structural characteristics of the finite-state language but, in terms of specific frequency of bigrams, was distinctly idiosyncratic.

The comparison between rank orders A and C is a reflection of the degree to which subjects are simply keying on the raw frequency data as displayed in the exemplars. The failure for this correlation to be different from zero suggests that subjects were not solving the anagrams based on superficial knowledge of frequency of bigrams nor upon a fixed set of memorized instances. They clearly acquired knowledge that can be characterized as deep, abstract, and representative of the structure inherent in the underlying invariance patterns of the stimulus environment.

This finding is analogous to the oft-cited Posner and Keele (1968, 1970) abstraction of prototype effect. The underlying prototypes that their subjects extracted from the exemplary dot patterns can only be specified by averaging the spatial relations among the various pattern components. But, psychologically, such an averaging is not just a simple piling up of the features of the exemplars. If memory behaved like that, the resulting representation would not be the distinct prototypes Posner and Keele found but a blob.

The induction routine that appears to be operating in situations such as these is necessarily one that results in an abstract representation. Moreover, it can be applicable to classifying novel instances and not specifically characterizable by raw compiling of experienced instances.

This issue is one of considerable complexity. The point of the preceding argument is not that all memorial systems must be viewed as being founded on induced abstractions. The evidence of Brooks (1978) and others (cf. Medin, 1989; Smith & Medin, 1981) shows that memories are frequently based on instantiations, fairly uninterpreted representations of the stimulus inputs. The point is that when implicit acquisition processes are operating, the resulting memorial system is abstract. As was shown elsewhere (Allen & Reber, 1980; Reber & Allen, 1978), the same subjects working with the same grammars can emerge from a learning session with either an instantiated memory system or an abstract one, depending on the learning procedures used.

In those papers that old war horse "functionalism" was shown to provide the best characterization of this issue. That is, the specific functions that need be carried out invite the learner to assume a cognitive stance that is functional, that will accomplish the task at hand. Under some circumstances (such as the paired-associates learning procedure used by Brooks, 1978, and Reber & Allen, 1978) a rather "concrete," instantiated memorial system will be established; under others (the instant asymptote technique in the PL studies [Reber & Millward, 1968] and the observation procedure in Reber & Allen, 1978) a distinctly abstract representation will emerge. In the implicit, unconscious acquisition mode, the default position appears to be abstraction.

On mental representation

As Rosch and Lloyd (1978) pointed out, sooner or later every discourse on mental process and structure must come to grips with the problem of the form of the representation of knowledge held. Such discussions must begin with some presumptions. The ones introduced here are, of course, open to emendation as understanding progresses. They are taken as the starting point simply because they have considerable explanatory power, more than most contemporary cognitivists have granted.

First, the general arguments put forward by such diverse theorists as Gibson (1966, 1979), Garner (1974, 1978), and Neisser (1976) that the stimulus is more than the physical setting for the occurrence of a response is taken as a given. This point is more than a simplistic swipe at behaviorism; it is an argument that stresses that the stimulus domain within which we function is extraordinarily rich and complex and, in all likelihood, much more so than most cognitivists have been willing to recognize. The underlying causative nature of the stimulus environment is rarely explored; most theorists are satisfied with characterizations that are theory driven.

Second, there is general agreement with the arguments put forward (in rather different forms, to be sure) by Palmer (1978) and Anderson (1978, 1979, 1983) to the effect that most theoretical attempts to deal with the representation issue are misguided. Palmer maintains that the confusion derives from a failure to deal directly with metatheoretical factors concerning existing models. Anderson argues that, in principle, there are no ways in which behavioral data can be used to identify uniquely any one particular mental representation. There are some reasons for perhaps disputing these claims (see, e.g., Hayes-Roth, 1979; Pylyshyn, 1979, 1980), but they are not a concern here.

From the point of view taken as presumptive here, it matters not at all whether the following interpretations of mental representation are supported by a well-structured consideration of metarepresentational factors or whether they can be shown to be uniquely specifiable. The point of view taken here is reflective of that of classical functionalism as introduced in the preceding section. Functional theories are typically regarded these days as formulations, abstract to be sure, of what it is possible for a person to process and why. This approach seems right, and as has been argued elsewhere (Allen & Reber, 1980; Reber, 1989a; Reber & Allen, 1978), the main consideration should be with characterizing representations in terms of how the individual can be seen as behaving in an adaptive fashion rather than in terms of pure representational theory. For example, as previously discussed, there are good empirical reasons for regarding the functional representation of the mental content of a finite-state grammar as an ordered set of bigrams (or perhaps larger chunks, see Mathews, Buss, Stanley, Blanchard-Fields, Cho, & Druhan, 1987a, and Servan-Schreiber & Anderson, 1990) and not as a formal Markovian system.

Third, the oft-dismissed position of representational realism is accepted as a first approximation. What is in the stimulus world is what ends up in the mind of the perceiver/cognizer. The point here is that a good starting-off place in dealing with the representation problem is the physical stimulus itself. Under various constraints of processing and various task demands, enrichment and elaborative operations are certainly employed, and the resulting coded representation may very well not be isomorphic with the stimulus field. Nevertheless, as Mace (1974) put it, sometimes a good initial strategy is to "ask not what's inside your head, ask what your head's inside of."

Several findings from studies with artificial grammars are relevant to the issue of representation. Table 2.2 gives the summary data from 14 separate experiments run in our laboratory that reveal some interesting patterns. Some details on procedure may help to understand this analysis. In all of these studies, the knowledge acquired during learning was assessed using the standard well-formedness task in which subjects are presented with a number of test strings (typically 100) that must be classified as either grammatical or nongrammatical. In the typical experiment, the 100 trials consist of 50 unique items, each of which is presented twice without feedback about the correctness of the response.

This procedure yields data that speak directly to the representational issue. The logic of the analysis is simple. There are four possible outcomes for each individual item for each subject. The subject may classify it correctly on both presentations (CC), classify it correctly on only one of the two (CE or EC), or misclassify it on both presentations (EE). Assume that the subject operates using a simple decision-making strategy. When the status of the item is known it is always classified correctly; when it is not known, a guess is made. This simple model is quite powerful and allows for a surprisingly deep analysis of the representation problem. Specifically, under this model:

1. The values of CE, EC, and EE should be statistically indistinguishable from each other, and all should be significantly lower than the value of

Table 2.2 Correct (C) and Error (E) response patterns to individual items during well-formedness tasks: Summary table from 14 experiments with artificial grammars. As noted, each of the experiments used a particular training procedure during the acquisition phase.

	Pattern				
Training Procedure Used	CC	CE	EC	EE	Consistency
1. Simple memorization of exemplars	.69	.07	.12	.12	.81
2. Simple memorization of exemplars	.66	.10	.11	.13	.79
3. Memorization of exemplars with explicit instructions	.53	.12	.12	.23*	.76
4. Simple observation of exemplars	.73	.08	.09	.11	.84
5. Paired-associates learning	.65	.12	.07	.16*	.81
6. Random display of exemplars with implicit instruction	.51	.16	.14	.19	.70
7. Random display of exemplars with explicit instructions	.48	.12	.14	.25*	.73
8. Structured display of exemplars with implicit instructions	.52	.16	.16	.16	.68
9. Structured display of exemplars with explicit instructions	.68	.10	.10	.11	.79
10. Observation of exemplars with explicit rules at beginning of observation	.67	.11	.12	.11	.78
11. Observation of exemplars with explicit rules in middle of observation	.58	.12	.14	.16	.74
12. Observation of exemplars with explicit rules at end of observation	.57	.13	.15	.16	.73
13. Explicit rules only	.54	.11	.16	.18*	.72
14. Observation only	.48	.15	.13	.24*	.72

*Cases where the EE value was significantly higher than the mean of the CE and EC values.

A. S. Reber, 1989, *Journal of Experimental Psychology: General, 118*, 219–235. Copyright 1989 by the American Psychological Association.

CC. This pattern is expected on the grounds that the items that contribute to CE, EC, and EE are those about which the subject's knowledge base is not relevant.

2. A value of EE significantly greater than the values of EC and CE is prima facie evidence for the elaboration of nonrepresentative rules on the part of subjects. The point is if subjects emerge from the learning phase with rules (either explicit or implicit) that are not accurate reflections of the grammar, this knowledge base will consistently lead them to misclassify particular items.
3. The robustness of representative knowledge can be assessed by the relationship between the values of EC and CE. If the value of CE is detectably larger than EC, we can reasonably suspect that forgetting was occurring

during testing, and correspondingly, if EC is larger than CE, we can infer that learning was occurring during testing.

4. An estimate of knowledge of the grammar can be made by looking at the value of CC, which contains only those items whose status was known by the subjects plus those guessed correctly on both presentations.
5. An overall measure of consistency of responding can be derived by taking the sum of CC and EE.

Table 2.2 gives these several values from 14 different experimental conditions. The uninteresting ones can be dispensed with first. There are no cases where the values of CE and EC are significantly different from each other. Thus, there is no evidence of loss of knowledge during the well-formedness task and no evidence of any additional learning taking place.

The interesting results are those concerning comparisons between the values of EE and those of EC and CE. When no difference is found between EE and the mean of CE and EC, it is reasonable to conclude that there is no evidence of nonrepresentative rules in use. Values of EE that are large relative to those of EC and CE can be taken as evidence that subjects were using rules that are not representative of the grammar. The starred values in Table 2.2 are the five conditions that yield evidence of subjects emerging from the learning phase with notions about structure that were not commensurate with the stimulus display.

It is instructive to look closely at these five cases. In one (Condition 13), the subjects were only given the schematic diagram of the grammar but no opportunity to observe exemplars. It would appear that, not surprisingly, such a procedure encourages subjects to invent specific rules for letter order and, in the absence of complete learning, to elaborate rules about permissible letter sequences that are not reflective of the grammar.

Two of the conditions (3, 7) are illustrative of what happens when subjects are under an instructional set that encourages the use of rule-search strategies but where the letter strings are given to them in a haphazard order. Such a set of demand characteristics encourages subjects to invent a sufficient number of inappropriate rules to inflate the EE values.

Condition 5 used a paired-associate task to impart knowledge. As we argued in the papers that yielded these data (Allen & Reber, 1980; Reber & Allen, 1978), the very nature of the PA task leads subjects to set up an instantiated memorial system composed of parts of items and some whole items along with their associated responses. Hence, the inflated EE value is not due to the application of inappropriate rules; rather, it is due to subjects' tendency to misclassify test strings because inappropriate analogies exist in instantiated memory. The final condition with an inordinately high EE value is Condition 14. There is no obvious explanation for this outcome. This datum is an anomaly; one such outcome out of fourteen, however, is not bad at all.

The remaining nine conditions all yield response patterns that fit with the proposition that whatever subjects are acquiring from the training sessions can be viewed as basically representative of the underlying structure of the stimulus domains. These consist of "neutral-set" conditions, in which the subjects are led to approach the learning task as an experiment in memory or perception and no mention is made of the rule-governed nature of the stimuli (Conditions 1, 2, 4, 6, 8), and of "structured-set" conditions, in which subjects are provided with information concerning rules for letter order but in a manner that ensures that conscious rule searching will be coordinate with the kinds of rules in use (Conditions 9, 10, 11, 12).

Taken together, these experiments lend general support to the proposition that implicit learning functions by the induction of an underlying representation that mirrors the structure intrinsic to the environment. Such an induction process takes place naturally when one is simply attending in an unbiased manner to the patterns of variation in the environment or when one is provided with an orientation that is coordinate with these variations.

This characterization of the appropriateness of mental representation tells us nothing about the sheer amount of knowledge that one takes out of a learning session. In fact, it is relatively easy to show that there is little to choose between explicit and implicit processes here. The consistency values in Table 2.2 reveal surprisingly little variation from condition to condition, particularly when compared with the range of CC and EE values. These consistency values can be seen as a kind of "raw" estimate of the total number of rules that subjects can be said to be using during decision making for they are simply the sums of the CC and EE values. Taking a simple (and only quasi-legitimate) average across conditions reveals that the overall mean consistency values for the starred conditions and the nonstarred ones are .75 and .76, respectively. Thus, there is no evidence that either set of conditions produces more rule learning; the difference is that explicit learning frequently results in the emergence of a number of inappropriate rules while implicit learning tends to yield representative, veridical rules.

This same model of representation is supported by data from other tasks. In the Reber and Lewis (1977) anagram solution task, subjects worked with the same problem sets over four days. In this study cues were provided on some trials, and as a result, there was improvement over time. To adjust for this increase in the probability of a correct response, a stochastic model (details on which can be found in the original paper) was fit to the data and used to predict the pattern of error and correct responses to individual test items that would be expected under the assumption that subjects were not using inappropriate rules. The results were in keeping with the general theme here. The EEEE value (the proportion of items solved incorrectly on all four days of the study) was no higher than would be expected under the assumption that subjects either knew the solution to a particular anagram or made nonsystematic guesses for problems not within the domain of their knowledge base.

Several experiments using the probability-learning procedure are also of interest. The relevant data are the recency curves. In the standard analysis of a PL experiment, a recency curve presents the probability of a given response plotted against the length of the immediately preceding run of that event. Recency curves may take on any of a number of shapes depending on the conditions of the experiment. Negative recency occurs most commonly, particularly early in an experiment. Under some rather special circumstances, however, even positive recency may be observed (see Friedman, Burke, Cole, Keller, Millward, & Estes, 1964, for details). The concern here is with the recency curves from experiments with 500 or more trials having a Bernoulli event sequence with the probability of the more likely event set at .80. That is, experiments with "ordinary" event sequences and with a sufficiently large number of trials so that short-term vagaries are averaged out.

Figure 2.9 presents the pooled recency data from five such experiments (see Reber, 1967b; Reber & Millward, 1968). All subjects were run through a learning period using either traditional PL or the instant asymptote technique. The dashed curve gives the overall probability of a subject predicting the more frequent event given that the preceding run of that event is of length k. The solid curve gives the comparable statistic for the event sequences actually used in these experiments. The subject generated curve has been adjusted downward by exactly .04 at all points to correct for a ubiquitous overshooting effect that is observed in all of these many-trial experiments (see Reber & Millward, 1968,

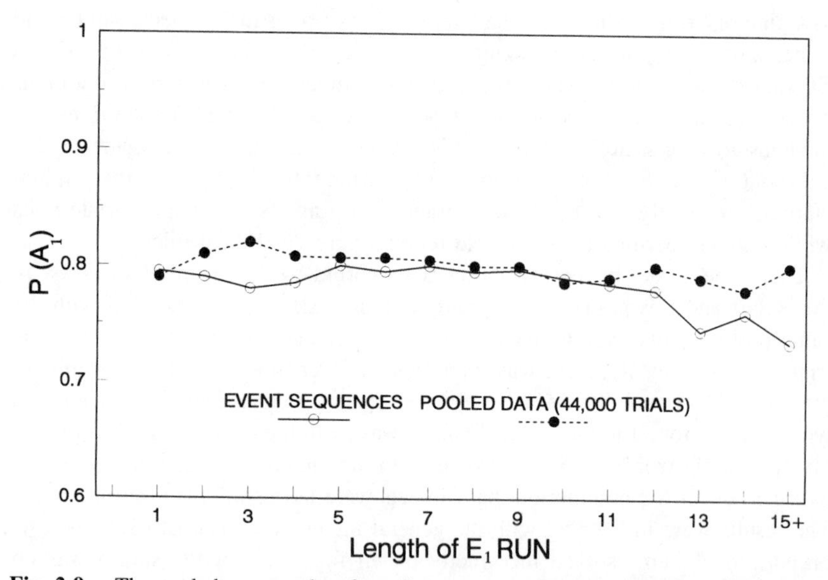

Fig. 2.9. The pooled recency data from five separate probability learning experiments. Open circles give the average proportion with which the more frequent event actually occurred over all event sequences used; filled circles give the average proportion with which subjects actually predicted the more frequent event. Note that the subjects' curve has been adjusted by .04 to correct for overshooting. These data are based on 44,000 trials from 88 asymptotic subjects run with the probability of the more frequent event set at .80. This figure is adapted from "Implicit learning and tacit knowledge" by A. S. Reber, 1989, *Journal of Experimental Psychology: General, 118*, p. 229. Copyright 1989 by the American Psychological Association.

for a discussion of this issue). This adjustment in no way modifies the startling aspect of these two curves.

With few exceptions, the curves sit on top of each other. There is no evidence whatsoever for either the positive recency predicted by the early conditioning models or the negative recency reported by many. There is, however, overwhelming evidence for a mental representation that reflects the structure of the stimulus environment. The simplest characterization of this curve, which is based on a total of 44,000 responses, is that it reveals subjects mimicking the structure of the event sequence. Subjects' prediction responses show flat recency curves because the event sequences themselves display flat recency curves—as they must, being Bernoulli in nature.

This is not a new point, having been made by both Derks (1963) and by Jones and Meyers (1966), who showed that either positive or negative recency can be encouraged by presenting event sequences with either many long or many short runs of events. But the precision with which subjects' response patterns can reflect the event patterns has never really been appreciated. To take this point to a further extreme, data like those in Figure 2.9 are so robust they can actually be used as a check on one's experimental procedure. In one of our early PL studies (Millward & Reber, 1972), the subjects' overall response proportions were .523 and .476 for the two events, a result that was perplexing since each event had been programmed to occur on exactly half of the trials. The anomaly turned out to be in the computer program used to generate the sequences. A check revealed that the two events had actually been presented to subjects with proportions of .520 and .480!

Although the preceding analyses seem to provide support for the representational realist position, it is still unclear just how far one can legitimately push such a proposition. In many of the experiments reported here and in other related areas of study (see, for example, Schacter's 1987 article outlining how the notion of implicit memory has been used in cognitive psychology), subjects respond in ways that indicate that their mental "content" may not be quite so neatly isomorphic with that of the stimulus field. However, it also seems reasonably clear that when such transforms or constructions of representations are observed, "secondary" processes are responsible. That is, the "primary" process of veridical representation of environmental structure becomes colored by either elaborative operations, as in experiments where instructional sets encouraged invention of inappropriate rules (Howard & Ballas, 1980; Reber, 1976), or restrictive operations in which task demands led to the narrowing of attentional focus (Brooks, 1978; Cantor, 1980; Reber et al., 1980).

Moreover, I would be remiss if I did not add that careful scrutiny of the EE values in Table 2.2 reveals that all is not so neat as one might hope. It is quite noticeable that even in the nonstarred conditions there was a tendency for some nonrepresentational elaboration to take place. In all nine of these cases the EE value is equal to or higher than the EC or CE values. This value is significant even by the weakest of statistical tests, the sign test, where $p < .05$. Moreover, in a recent paper comparing connectionist and memory-array models of implicit learning, Dienes (1992) used as the data base against which the models were compared one that fixed the EE value as greater than the average of CE and EC by .05.

In summary, the problem of mental representation is clearly no easy nut to crack. The position taken here seems to be a reasonable one, and temporarily taking on the role of a good Popperian, I acknowledge that it will probably be shown to be wrong after analysis. For the moment at least, the data support the interpretation that tacit knowledge is a reasonably veridical, partial isomorphism of the structural patterns of relational invariances that the environment displays. It is reasonably veridical in that it reflects, with considerable accuracy, the stimulus invariances displayed in the environment. It is partial in that not all patterns become part of tacit knowledge. It is structural in that the patterns are manifestations of abstract generative rules for symbol ordering. I would like to stay with this position until, using a classic Popperian falsificationist model, the data show it to be indefensible.

Now admittedly, there are a number of issues lurking underneath this simple gloss. They concern such problems as: (1) determining the degree to which the functional stimulus processed by the subjects can be reasonably characterized as having the same formal structure as the display the experimenter believes he or she has constructed; (2) attempting to establish whether the subject's tacit knowledge base can be said to correspond, in a formal way, to the structure of the display; and (3) establishing a formal characterization of the structure of the display and determining the degree to which it corresponds with the intended structure. In Chapter 4, in the sections entitled "On rules" and "Knowledge representation," these and related issues are discussed at some length.

On the availability of tacit knowledge

The conclusion reached in the very first of the studies on implicit learning (Reber, 1965) was that the knowledge acquired was completely unavailable to consciousness. The many experiments carried out since have shown that position to have been an oversimplification. The picture that is emerging, while perhaps somewhat less striking, is certainly more interesting. Specifically, knowledge acquired from implicit learning procedures is knowledge that, in some "raw" fashion, is always ahead of the capability of its possessor to explicate it. Hence, while it is misleading to argue that implicitly acquired knowledge is completely unconscious, it is not misleading to argue that the implicitly acquired epistemic contents of mind are always richer and more sophisticated than that which can be explicated.

In Reber and Lewis (1977) data were first presented to support this position. Over the four days of that study, during which subjects solved anagram puzzles based on the syntax of an artificial grammar, there was a general increase in the ability of subjects to communicate their knowledge of the rule system in use. There was also an increase in the ability to solve the anagrams—but the former never caught up with the latter. That is, as subjects improved in their ability to verbalize the rules they were using, they also improved in the richness and complexity of the rules they were acquiring and using. Implicit knowledge remained ahead of explicit knowledge.

In a recent study, Mathews, Buss, Stanley, Blanchard-Fields, Cho, and Druhan (1989) employed a novel variation on the classic yoked-control technique to explore this issue. In an artificial-grammar learning experiment, subjects were interrupted at intervals during the well-formedness judgment task and asked to explicate the rules they were using to make their decisions. Transcripts were made of these explications and then given to yoked-control subjects, who were then run through the same well-formedness task. So equipped, these control subjects, who had no learning experience, managed to perform at roughly half the level of accuracy of the experimental subjects. Moreover, the later in the experiment the transcripts were taken, the better the yoked controls were, but the gap between them still remained.

In a recent study, Stanley et al. (1989) reported one of the more tantalizing findings in this area. The experiments in this paper were run using the Broadbent control procedure and the yoked-control technique. The experimental subjects were tracked carefully over many days, and verbal statements were taken from them on a regular basis. As they progressed through the experiment, their performance levels increased as did their explicit knowledge of the rules of the systems. However, the two curves, so to speak, followed different functions. Subjects' overall level of performance on the task reached asymptote relatively early in the study and seemed to occur rather suddenly. But this increase in performance was not accompanied by any increase in the ability to explicate this knowledge. Reasonably accurate explication of the rules was not observed until the very final trial blocks of the experiment, long after subjects' behavior indicated that they had induced the requisite knowledge base to succeed in the task.

This finding has a familiar quality to it. It is very much like the kinds of introspective reports that one gets from expert chess players (DeGroot, 1965). They frequently play through an extremely complex and subtle line without being consciously aware of the nature of the plan. They often report only a vague sense of the "rightness" of this line of play with little concrete, verbalizable basis for it. Later, after the game, they will engage in extended "postmortem" sessions where they attempt to make explicit the nature of the line of play. Here, too, the implicit base that is effectively directing behavior is "way ahead" of the explicit instantiation of it.

This line of thinking has, needless to say, not gone uncontested. Dulany has argued that this gulf between tacit knowledge and knowledge that can be consciously expressed is not nearly so great as all this. In an attempt to deal directly with this issue, he and his colleagues carried out a replication of the standard grammar learning task but introduced some interesting variations on how subjects' conscious knowledge was assessed (Dulany et al., 1984).

The experiment followed the standard design with a learning phase during which subjects memorized letter strings and a testing phase during which they had to determine the well-formedness of novel strings. They introduced an innovation during the well-formedness task that required subjects to specify the feature or features of each test item that led them to classify it as they did. That is, for items they felt were acceptable, they were to underline the letter or letters they felt made that item grammatical; for items they felt were not acceptable, they were to cross out the letters that rendered the string ungrammatical.

Dulany et al. estimated the validity of each of the marked features by determining what a subject's level of performance would theoretically have been if they were, in fact, using these rules in a systematic fashion. They then compared each subject's projected performance level with their actual performance level based on the validity estimates. They found no residual leading them to the conclusion that subjects were, in fact, conscious of the rules they were using.

This is a rather ingenious technique but, as we have argued elsewhere (Reber, Allen, & Regan, 1985), it has some problems. First, the task confronting a subject who feels that a given string is grammatical is a rather peculiar one. For example, how would one deal with such a request if one were working with a corpus from a natural language? What, indeed, was there about the preceding sentence that made it grammatical? Why did subjects not simply underline the entire string each time?

Second, there are reasons for thinking that the nature of the task carried its own guarantee of success. That is, the task may work to "force" the data to appear as though they carried the implication of awareness even if the subjects were only reporting vague guesses about the appropriateness or inappropriateness of letter groups. For example, suppose that a subject is confronted with the string PTVPXVSP (which is, in fact, nongrammatical, see Figure 2.1) and experiences a vague sense that there is something about the end of it that makes it feel wrong. She then crosses out the last two or three letters. Dulany would take this to mean that the subject "knows" that a string like this may not end with the sequence VSP and make a projection about performance based on such a "rule." Now the projection may very well account for the subject's performance, but it certainly does not force the conclusion that the rule was held consciously. The problem lies not with the method but with the attribution, a point also made by Servan-Schreiber and Anderson (1990). In our reply to Dulany (Reber et al., 1985), we showed that a Monte Carlo type simulation using the Dulany technique can account for the results. Dulany et al. (1985) presented reasons for doubting our analysis. This particular issue will continue to be disputed.

Finally, the Dulany et al. procedure fails to distinguish between knowledge that is available to consciousness after attempts at retrieval and knowledge that is present in consciousness at the time the decisions themselves are being made. The paper by Stanley et al. (1989) previously discussed shows how complex this issue is likely to turn out to be. Carmody et al. (1984) have also noted this problem in assessing the knowledge base that physicians are taught to use versus that which they actually use in diagnosis. Furthermore, Schacter (1987) has argued that conclusions reached about the availability of implicit information must take account of a variety of task constraints that have their own impacts. It is one thing to be able to struggle to bring some mental content to consciousness "after the act" and another thing entirely to assume that conscious factors were playing a causal role "during the act."

Parallel arguments were made recently by Perruchet and Pacteau (1990). They maintained that subjects do not induce abstract representations of the artificial grammar but, rather, establish simple, fragmentary representations based on the patterns of bigrams that are displayed in the exemplars subjects studied during learning. To support this line of argument, they used two different learning techniques. In one, subjects scanned a sheet of paper on which was printed a list of 20 exemplars from an AG; in the other, subjects scanned a list of bigrams that were found in those exemplars adjusted for the frequency with which each was found in the full set. Perruchet and Pacteau found that both groups were able to distinguish well-formed from ill-formed strings and concluded that the our original conclusion that subjects induced an abstract representation of the stimulus display was unfounded. If, they argued, subjects could make well-formedness judgments at rates better than chance when they were only exposed to the bigrams of the full strings, then it was likely that their representations were fragmentary.

There are many problems with this study that were pointed out by myself (Reber, 1990) and Mathews (1990, 1991). The primary ones are that (1) the learning procedure used, in which subjects are given both the stimuli printed out on a sheet of paper and time to freely peruse the list, clearly invites the establishment of instantiated memories; (2) their demonstration is, at best, indirect in that the finding that subjects given only bigrams performed better than chance does not entail the conclusion that the control subjects given the full strings were encoding them as bigrams; (3) although the subjects given only the bigrams performed above chance on the well-formedness task, they also performed significantly poorer than those presented with the full strings, suggesting that the fragmentary representation theory does not account for the full range of performance; and (4) the fragmentary argument misses the fact that spatial information is captured by the stimulus displays and that this represents an abstract encoding that can be captured by some of the more sophisticated models currently being entertained (e.g., the THYIOS model of Mathews and his co-workers, Mathews, Druhan, & Roussel, 1989).

A rebuttal by Perruchet and Pacteau (1991) introduced a number of issues that relate to the general question of what kinds of representations subjects can be said to emerge from implicit learning experiments with. The thrust of their argument is a most fascinating one. They clearly feel that there are occasions when cognitive processes take place that are outside of the scanning and monitoring functions of consciousness. Similarly, they feel that there are occasions when mental representations are established that can only be regarded as abstract in nature. However, they specifically exclude the possible of the co-occurrence of an abstract, unconscious process. These issues go beyond the focus of this chapter, which is to try to review the data for and against the standard view of implicit learning. In Chapter 4, under the headings "Knowledge Representation" and "On Consciousness," these issues are put in an philosophical context and discussed in considerable detail.

Methodological issues in implicit and explicit learning

For the purpose of the issues raised by the research under discussion, I plan to try to avoid getting embroiled in the neighboring topics of semantic and affective activation without awareness (see Holender, 1986) and the larger questions of subception and subliminal perception (see Erdelyi, 1974, 1985). However, we need to take at least a glancing look at them because they have methodological lessons to offer, lessons that help untangle some the problems encountered in the preceding sections. In the next chapter I will have (a little) more to say about these topics.

Whenever one confronts an issue like the implicit/explicit distinction there is a tendency to "either-or" the problem. I referred to this in an earlier section as the *polarity fallacy*. In some considerable measure, the arguments over whether or not the epistemic contents that emerge from experiments on implicit learning can be shown to have conscious components result from the participants being seduced, as it were, by the fallacy. The thrust of the Dulany et al. position and similar arguments recently proposed by Perruchet and Pacteau (1990; 1991) is that all that is abstract is conscious, that the entire fabric of data from the typical artificial-grammar learning study can be understood from the point of view that subjects are either (1) aware of the detailed abstract nature of the rules they are using to make decisions or (2) the knowledge base is concrete and fragmentary and not abstract. This is why they, like Dulany et al. (1984, 1985), go to such lengths to show that the projections made from the subjects' responses account for all of the data without residual.

In a parallel vein, Brody (1989) has argued that the data collected and examined in the whole body of research on implicit learning is suspect because nowhere in it were the "proper" tests run to determine whether subjects had (any) conscious knowledge. Brody's point, in essence, is that we should not acknowledge the reality of implicit processes or tacit knowledge until all tests for explicit processes and conscious knowledge have been carried out. Dulany's position, which fits comfortably with Brody's, is that we should always favor consciousness as the explanatory operating principle until we show conclusively that it cannot account for all the data. These two, mutually supportive positions share a common presumption, that consciousness takes epistemic priority.

The overt issues here are, at their core, essentially ones of methodology and measurement (although it will be seen that they are merely the empirical focus for deeper epistemological issues). They were first raised by Eriksen (1958) over the question of subliminal perception and reintroduced recently in several quarters around the issues of unconscious semantic activation (Holender, 1986) and implicit learning (Brody, 1989; Dulany et al., 1984, 1985). They revolve around the question of whether the tests for awareness that have been used in the various studies cited above were sufficient to support the conclusion that the learning was unconscious and the resulting knowledge base was tacit.

Eriksen was right to introduce the methodological issues in the form that he did. Researchers on subception had been confusing perceptual processes with response processes; when they thought they were measuring a failure of conscious perception of, say, a "dirty word," they were, in fact, measuring a reluctance on the part of the subject to respond with that "dirty word." However, while the Eriksenian position here was good counsel and can be seen in the contemporary criticisms of Holender (1986), it, too, has problems. This position, common though it is in current psychological thinking, is not the optimum one to take. It suffers from an unusual variety of methodological sophistry, what Erdelyi (1986) called the problem of *experimental or methodological indeterminacy.*

As I pointed out in Chapter 1, in order to demonstrate tacit knowledge for any specified stimulus domain, the relationship between the implicit and the explicit can be reduced to a simple inequality

$$\alpha > \beta$$

where α is the sum of information available to the unconscious and β is that available for conscious expression. The circumstances claimed as optimal by Eriksen, Brody, and others for establishing the existence of implicit knowledge are those in which $\beta = 0$ and $\alpha > 0$. This relationship pertains when all tests for awareness yield estimates of β statistically indistinguishable from 0 and behavior is displayed that could only have been caused by α. It is this position that I regard as untenable—first because it puts the burden on the wrong party and second because I can envision a kind of Popperian nightmare where no possible evidence could be provided that could clinch the case for the $\beta = 0$.

This position emerges from a particular epistemology, a point of view, for want of a better term, we can call the *consciousness stance*. From this perspective, consciousness takes priority; awareness and self-reflection become the central features of human cognitive function. The cognitive unconsciousness is dealt with by exclusion and implicit processes have to be defended against claims of residual awareness. The popularity that this consciousness stance has in contemporary psychology is in no small way due to the very poignant sense that awareness and sense of self are what defines our very existence and, therefore, are the features that make us human. George Miller recently called consciousness "the constitutive problem of psychology" (Baars, 1986, p. 220). The stance that gives the unconscious and the implicit priority is one that, one might argue, strips away something fundamentally human from our characterization of ourselves.

Yet, as I will argue in the next chapter, there are good reasons for endowing the unconscious and implicit systems with cognitive priority—for taking the, for want of a better term, *implicit stance* ("unconsciousness stance" has an unhappy ambiguity, indeed it is almost an oxymoron). Accordingly, there are good reasons for arguing that the burden of proof belongs, not on those who maintain that a given cognitive function is implicit in nature, but on those who would argue that it is conscious and explicit.[7] The arguments that have been put forward by Brody, Holender, and others are of the other kind; their position is that the workers in the field of implicit learning and memory have not demonstrated convincingly that $\beta = 0$ and that we should resist the conclusion that implicit processes are operating.

To attempt to achieve this ideal state is a mistake. Moreover, it misses the point of the research, which is not to show that consciousness is totally absent from the process but instead that explanations of behavior based solely on conscious factors do not provide satisfactory accounts. And, of course, there are the serious problems of methodology and measurement that accompany the task of estimating the values of α and β. Simply, no matter which stance one takes, there will always be the psychometric problems associated with the measurement of mental content. It is going to be hard enough to determine when we can conclude with assuredness that $\alpha > \beta$ without insisting that the value of β be null. To make this point as clearly as possible imagine the following scenario:

In a well run, carefully controlled study with meaningless geometric forms as the stimuli, researchers have shown that affective processes have a lower threshold for activation than recognition processes. That is, subjects show a marked preference for geometric forms that they were exposed to in an earlier phase of the experiment over novel forms; although when asked to select which of the forms they had, in fact, seen, they are at chance levels of performance (see Kunst-Wilson & Zajonc, 1980; Seamon, Brody, & Kauff, 1983; Seamon, Marsh, & Brody, 1984). In these studies the researchers concluded that $\beta = 0$ because the two-alternative, forced-choice (2AFC) procedure used for the recognition task yielded chance results. The 2AFC procedure is, to date, the most sensitive test we have for the capacity to discriminate and in these contexts is generally taken as the strongest test for awareness.

However, it is quite simple to argue that even here it is perfectly plausible that $\beta > 0$, that subjects actually had some conscious content that described the geometric forms but, just like in the old "new look" days (see Erdelyi, 1974), refused to use it. Consider a subject in the recognition phase of this study confronted with the question, "Which one of these items did you see before?" Suppose that the subject in this experiment does not like being wrong, particularly when she has made a conscious effort to be right. On the other hand, she does not terribly mind being wrong, not because she tried and failed but simply didn't try. She is confronted with a task that clearly has right and wrong answers, for the instructions indicate that one of these geometric forms has, in fact, been part of an earlier display. She now ignores the small piece of conscious knowledge she has about the stimulus because it is not enough information for her to use and risk being wrong. Instead, she flips a mental coin and guesses blindly, yield-

[7] Interestingly, Dulany and his co-workers appreciate this point and in their papers do indeed go to considerable lengths to show that conscious processes can account for observed behavior. This is one of the reasons why the dispute between us is such an interesting one.

ing, of course, chance performance, but the experimenter concludes that $\beta = 0$. Now, consider the subject in the other context where she is confronted with the question, "Which one of these two items do you prefer?" Here the pressure of being wrong is not present; there are no right and wrong answers, only personal preferences. Now, she feels perfectly comfortable calling upon that tiny amount of conscious knowledge of the stimulus and uses it to select the previously exposed item—because, as we know, familiarity breeds affection.

Under such a scenario even the most psychometrically sound measure of awareness comes up short. It is, moreover, not a terribly unlikely scene and could very well be operating in these experiments. Were one to invoke this characterization of these studies one would call into question the conclusion that so many have reached—that they are genuine reflections of unconscious processes.

The point here, of course, is not to dismiss the notion that unconscious processes are operating in these experiments, quite the contrary. The strong entailment of this little *gedenken* experiment is that the way in which empirical questions about awareness are asked leads to different estimates of α and β. It is simply not possible to do all possible tests for consciousness or, to put it even more strongly, for every indicator that may provide an estimate of $\beta = 0$, there is surely another that would yield an estimate of $\beta > 0$. The position that has been argued for since Eriksen allows no escape from methodological indeterminacies. One can never establish the pure state that would be required to show that awareness was completely lacking as a causative factor.

Moreover, independent of this futile search for the conditions under which $\beta = 0$, there are other problems involved in estimating the true values of α and β. As Jacoby and his coworkers have pointed out, in the study of implicit memory (where this issue of trying to set β at 0 seems to be somewhat less intense than it is in the study of implicit learning) there are likely cases where the measures of α are *underestimates* of the amount of knowledge that is actually held unconsciously, owing to the fact that the measures typically used to assess explicit memorial content are contaminated by hidden contributions from implicit representations (Jacoby, 1992; Jacoby, Lindsay, & Toth, 1992; Jennings & Jacoby, 1992). Jacoby (1992) examined the relationship between implicit and explicit memory using the *task dissociation* technique. In addition to using the two standard procedures, one to measure explicit memory (like free recall of previously presented words) and one to measure implicit memory (like stem-completion), he introduced a third, dissociation procedure. Here subjects are given the stem-completion task but are instructed to make sure that they only use words that were *not* in the originally presented list. The argument here is simple: if free recall can be taken as an accurate measure of explicit memory, then subjects should be able to carry out this task perfectly and never use an item from the original stimulus set. However, under several different conditions Jacoby and his coworkers have found that items from the original list leak in and "contaminate" this measure. Hence, the implication is that using procedures like free recall to provide estimates of β have consistently yielded overestimates, since implicit memories are slipping by undetected by consciousness.

Clearly, these methodological issues are not going to be easily resolved; indeed, if we take Erdelyi (1986) seriously, there can be no resolution. My suggestion here is fairly straightforward if a bit radical.

The solution is to adopt the *implicit stance*. Here the role of unconscious processes is taken as axiomatic and one is relieved of the burden of showing that $\beta = 0$. There is still a clear and compelling methodological and psychometric obligation to show that $\alpha > \beta$, but that is an obligation that has always been assumed by the researchers in the field. In our work we have frequently acknowledged that our subjects have some conscious awareness of rules. We have pointed out on several occasions (Allen & Reber, 1980; Reber & Allen, 1978; Reber & Lewis, 1977; Reber et al., 1980; Reber et al., 1985), as have others (Berry & Broadbent, 1984, 1987; Mathews et al., 1988; Reingold & Merikle, 1988; Servan-Schreiber & Anderson, 1990; Stanley et al., 1989) that the critical feature is not that subjects be totally unaware of the underlying structure of the stimuli but that their conscious knowledge be insufficient to account for their behavior.

The arguments to be developed in the next chapter are wrapped around evolutionary considerations, and as will be seen, one of the positions to be argued is that consciousness should be viewed as a kind of executive process interpenetrating other cortical (and maybe some subcortical) functions and providing some measure of volitional control over action. As such, consciousness cannot be separated from the rest of the cognitive domain; to search for a case where $\beta = 0$ is to search for a unicorn.

3. Evolutionary considerations: the primacy of the implicit

This chapter has several parts.[1] The first part consists of an extended series of "introductory remarks" on a variety of topics. These discussions will set the stage for the later sections where the lines of argument hinted at are developed in more detail. As will become apparent, the early sections are preparations for the development of the model of implicit cognitive processes derived from considerations of evolutionary biology.

Some introductory remarks

If the preceding was the "empirical chapter," this one should probably be tagged as the "sensible speculation chapter." The speculation will surround the introduction of a theoretical framework for the examination of implicit learning and tacit knowledge based on a number of proposals from evolutionary biology. Although, as will become clear, the model is grounded on formal principles of ontogeny and phylogeny, I still tend to think of it as somewhat "speculative" in nature because to date there has been insufficient empirical work directed at these issues to convince a hidebound skeptic of the legitimacy of these proposals. It will be seen that while the evolutionary perspective gives an excellent account of much of the existing data, still, it makes several predictions that have little or no empirical support to date.

On the other hand, I think of it as "sensible" speculation for a number of reasons. First, the adaptationist, functionalist line of argument to be developed will be derived from a number of standard considerations within evolutionary biology. The proposals introduced will be shown to follow naturally from a line of thinking that extends from von Baer's proposals on embryology made during the 1820s to the recent formalizations of a theory of development of William Wimsatt and his colleagues at the University of Chicago (Schank & Wimsatt, 1987; Wimsatt, 1986, in press). As such, the "speculations" will turn out to be natural entailments of a well-structured model. Second, even though, as noted, there has been little empirical work directed specifically at these adaptationist issues, there is enough of an empirical foundation to supply a strong sense of rightness to the direction of the work. As I will show, a good deal of work originally aimed at other problem areas in psychology will provide a data base

[1] Much of the material in this chapter was presented in various colloquia over the past several years; portions may also be found in Reber (1992a, 1992b).

that lends support to the functionalist position taken here. Finally, taking the evolutionary stance here leads to some very interesting and perhaps unexpected entailments concerning the cognitive unconscious and the manner in which its functions are displayed in real world behaviors.

Daniel Dennett, the well-known philosopher, is fond of referring to Darwin's theory of evolution as the "great engine of discovery." And so it is. And, although it has been long recognized for its seminal role in the various disciplines closely related to Darwin's field of "natural history," acknowledgement of its impact in psychology has tended to be more muted. This has been particularly true in recent years where both the role it has played historically and its potential for shaping psychological thought have gone largely (although as will become apparent, not completely) unappreciated. In earlier chapters, I hinted that the work on implicit learning will be best viewed within an evolutionary context; in this chapter I want to develop that theme in more detail and extend the coverage to include various other aspects of the cognitive unconscious.

The history of psychology has long been one of my favorite side pursuits and I agree with the position, espoused in various forms by such notables as George Miller (1962) and E. G. Boring (1950), that the emergence of psychology as a science cannot properly be understood without a deep appreciation of evolutionary theory. We are a species whose form, structure, and behavioral repertoire are, like all others, the product of a selectionist process, one that has singled out some functions and left countless unknown others on Nature's great slag heap. Recognition of this most fundamental of truths has been a guide for psychology for over a century. Yet in our present context, in the midst of the "cognitive revolution" (Baars, 1986), such considerations are often lacking. While animal psychologists, ethologists, behavior geneticists, and other practitioners in subfields that are historically more closely tied to "natural history" are generally cognizant of this history and often couch their theories within an evolutionary framework, modern cognitive scientists have largely ignored the selectionist's approach.

Now, on one hand, this is an omission with which it is possible to have considerable sympathy. There are several fairly obvious reasons why Darwinian thinking has been largely absent in the cognitive sciences, and although I will argue that it has been a mistake not to challenge them, it is easy to understand how they have operated in recent decades to dictate the shape and direction of the field.

Some reasons for the neglect of evolutionary principles in contemporary cognitive science

First, most modern cognitive psychologists have found themselves more concerned with problems of knowledge representation than knowledge acquisition. As I pointed out in the previous chapter, this concentration on the problem of knowledge representation led, perhaps unwittingly, to a neglect of the problem of learning. The place where the selectionist issues lurk is, of course, learning.

What gets learned and what does not? What kinds of stimuli do particular species key on and why do they neglect others? Why are some broad classes of associations formed with ease and others only with labored trial and error? These are the kinds of questions that elicit evolutionary analyses. Bypassing the acquisition problem for the representation problem largely bypasses the invitation to explore these lines of interpretation.

Second, the richly interdisciplinary aspect of contemporary cognitive science has provided it with a core structure that is linked with the study of artificial intelligence, often with a resulting neglect of natural intelligence. Indeed, technological concerns often seem to take precedence over the psychological problems that originally motivated them. Natural intelligence is the product of the process of natural selection; artificial intelligence is the product of the process of artificial selection. This is no mere play on words. When AI programs or expert systems "fail," loosely speaking, they do so for reasons very different from those that cause species to "fail." The selectionist pressures yielding an AI program that "succeeds," loosely speaking, will thus be of an entirely different kind from those yielding a species that "succeeds." The natural selectionist element has been largely outside the purview of the modern cognitivist.

It is worth pointing out that some AI theorists, particularly Herbert Simon, have taken note of evolutionary arguments. Early on, Simon (1962) noted that evolutionarily useful systems are virtually always hierarchical. Without such a structure, of course, evolution itself becomes seriously problematical (a point creationists consistently misunderstand) and, importantly, that which did evolve would be exceedingly fragile. Simon has consistently argued that artificial systems should follow this selectionist principle. In various ways the hierarchical principle will be adhered to in the adaptationist arguments developed below—particularly in the formalizations of a theory of development inspired by Wimsatt's work—although the "tree-structures" that such systems typically display will only be assumed implicitly.

There is an ironic touch here in that it is the seminal work of Simon and his long-time colleague Allen Newell that Glaser (1990) cites as being primarily responsible for the shift toward information-processing models and away from the exploration of learning and knowledge acquisition. If I understand the situation here, Simon's dual roles in the countervailing elements of this issue are actually not surprising. Simon has always been deeply sensitive to the underlying formalizations of evolutionary biology, particularly as these principles display strong parallels with the development of complex systems in other domains from the natural (economics) to the artificial (AI and computer simulation). However, his predilection has been to concentrate on the formalizations and not on the kinds of issues that evolutionary biologists would normally raise, such as issues of the *plausibility,* in evolutionary terms, of various theoretical models of form and function. Just how interesting this juxtaposition of perspectives actually is will become apparent later in this chapter with the introduction of Wimsatt's model of development based on his notion of "generative entrenchment." There the impact of Simon's early formalizations of the development of complex systems (Simon, 1962) on contemporary models of evolutionary biology will be obvious.

Third, the priority given to formal modeling has tended to distract workers from basic problems of function when these problems do not lend themselves to formal instantiation. Questions concerning the causal role of consciousness in problem solving or the impact of tacit knowledge on decision making are often put aside because of the inherent difficulty of introducing such concepts into a formalizable system. How, indeed, is one to instantiate intuition? How do you represent consciousness in a production system? Often one gets the sense that the success of a research program is judged by the elegance of its simulations independent of the insight into psychological process that the instantiation may or may not have.

As a short aside here, it's worth pointing out that similar kinds of "adjustment" in priorities have occurred in our field before, most recently with the mathematical-model approach to learning. For over twenty years, beginning with Estes's early work on stimulus sampling theory and Bush and Mosteller's linear model (see Estes, 1950, and Bush & Mosteller, 1951), mathematical models of learning followed a course that has some intriguing similarities with that displayed by contemporary information-processing models. In both cases the psychological processes that became incorporated into the theoretical discourses tended to be those that lent themselves to mathematical or computationally formal representation. Processes whose essential properties left them outside of the domain of the formalizations were largely ignored. A variety of concerns that should have been paramount to psychologists were often shunted aside because of the constraints imposed by the available mathematics or other formal instantiations.

The models developed for the probability learning experiments described in the previous chapter are classic instances of this misplaced priority. The conceptualizations of a subject in a PL experiment were based on the assumption that the essence of his or her behavior was an association between the stimulus circumstances present when a choice was made and the effect of the outcome event that followed that choice (for a retrospective on this work, see Neimark & Estes, 1967). Many of the variables subsequently shown to effect a subject's choices (see Chapter 2, pp. 38–42) were either neglected in the formal models or excused as local perturbations—not because they were not psychologically "real" but because they were factors that one simply could not incorporate into the

[2] It must be noted that I make these arguments from a position as a once and future practitioner of both the early mathematical-model approach and the contemporary connectionist approach and not as a petulant antiformalist. A glance at the several papers on probability learning that Richard Millward and I published during the 1960s and 1970s will reveal a distinct attempt to put our findings within the broad confines of Estes's original stimulus-sampling theory. Moreover, I am also enthusiastic about many of the formal representations of contemporary cognitive

models.[2] The formalisms came to dominate the psychology in much the same way as they do today in many of the research programs in the cognitive sciences. Sometimes it appears as if we find ourselves looking for the key under the lamppost not because that is where it will most likely be found but because that is where the light is.

Fourth, from Popper onward, evolutionary theory has been subjected to criticism on the grounds that it is, on one hand, nonfalsifiable and, on the other hand, "too easy." The former has, I hope, been finally put to rest by the work of Gould (see any of the essays in 1977a, but especially "Darwin's Untimely Burial"), Richards (1987), and others. Evolutionary theory makes predictions that are quite clearly falsifiable—although, like with astrophysics, the data base is, on occasion, not "experimental" in the standard sense. The latter is often raised as a problem on the grounds that evolutionary arguments become little more than sophisticated "just-so stories." This criticism has had more force than the former because of a lamentable tendency of many early evolutionists to "work backward," that is, to attempt to construct evolutionary scenarios that could account for existing forms and structures.

Just in passing, let me state that for the arguments developed in this chapter, neither of these standard criticisms hold. As will become apparent, the evolutionary model developed makes strong and unambiguous predictions that are empirically testable and that, in a few cases, have already been subjected to experimental examination. Moreover, some of these predictions are not obvious in the sense that they are not the products of any other current model of implicit learning and, in a few cases, anticipate findings that run counter to standard cognitive theory. Second, there has been no "working backward" here. The principles of evolutionary biology that have been brought to bear on this problem were derived from standard models of biology and then applied to psychological functioning. There has been no attempt to divine how cognitive processes might have evolved by looking for the adaptive aspects of present-day functions.

---

scientists, particularly the PDP models of Rumelhart and McClelland (1986), the induction routines of Holland, Holyoak, Nisbett and Thagard (1986), and the classifier system of Mathews and his co-workers (Druhan & Mathews, 1989; Mathews, Druhan, & Roussel, 1989), all of which appear as though they may be able to give a reasonable simulation of the abstraction of the underlying structure of our artificial grammars, and Anderson's various models from HAM to ACT* (Anderson, 1976, 1983; Anderson & Bower, 1973), which attempt to give formal expression to theories based on production systems. Moreover, in recent years in our laboratory we have begun the exploration of a subclass of PDP models known as Simple Recurrent Networks (Elman, 1990; Cleeremans, Servan-Schreiber, & McClelland, in press) as providing a legitimate characterization of some of the basic data from implicit learning experiments (Kushner, Cleeremans, & Reber, 1991).

It is surely true that when your theory is presented in formal fashion, as all these have been, that it can be subjected to firm and unambiguous test. I have no objections in principle to this approach to psychological theory; my concerns arise only when researchers become so preoccupied with their formalisms that they neglect the psychological processes they were developed to explain.

However, whatever force these four arguments may have, it should be clear that identifying the reasons for the contemporary cognitivist's neglect of evolutionary thinking does not really justify it. Dennett's quip is hardly controversial. Over the past century, evolutionary considerations have yielded more insights into psychological function than any other "great idea." During this period, Darwin's selectionist model has had a greater impact on psychological theory and research than any comparable theory. Moreover, this impact is easily visible in areas that relate directly to modern cognitive psychology. The early work of the Chicago Functionalists, primarily John Dewey and James R. Angell, from the turn of the century on into the 1930s was specifically focused on the adaptive role of mental action and the utility of what they liked to call in those days the "higher mental functions." Classic texts from James's *Principles of Psychology* (James, 1890) to Woodworth's two editions of *Experimental Psychology* (Woodworth, 1938; Woodworth & Schlosberg, 1954) couched their approach in an evolutionary context. "Mental life," as James liked to call the essential subject of psychology, was treated as the end product of a long selectionist process and needed to be viewed as such.

However, for the most part the cognitive revolution that emerged in the past couple of decades has occurred with little recognition of either the powerful heuristics of evolutionary biology or their explanatory potential. Historically, this is even more surprising since one of the more interesting elements of the recent enthusiasm over things cognitive is that the "old" problems that fascinated and haunted earlier workers were put center stage once again (Blumenthal, 1987).

One of the subthemes of this work, as hinted at above, is that a *functionalist* perspective is needed in cognitive psychology. The functionalism that I argue for, however, differs in important ways from other points of view that have used that label. First, it needs to be distinguished from the turn-of-the-century Chicago variety referred to above (which is usually capitalized), most notably in that it is not limited to explorations of consciousness, which was Angell's primary interest. The Chicago school certainly emphasized evolutionary issues, but its approach was too restrictive in the kinds of mental life it pursued, neglecting, as good Lockeans, the unconscious for the conscious.

Second, the version of functionalism I am arguing for should not be confused with the "hardware-independent" functionalism that has been put forward in an attempt to deal with some sticky metaphysical problems that emerge from considerations of what it means to have a particular mental state (see, e.g., Block, 1980). This approach, which has been particularly influential in contemporary theories of epistemology and artificial intelligence, takes little or no heed of Darwinian principles. If anything, it is anti-Darwinian in that some of its champions (e.g., Fodor, 1983) eschew any form of adaptationist thinking. The functionalism I defend is the evolutionary, adaptationist version that argues that structures and mechanisms, processes and behaviors need to be nested within an evolutionary framework and assessed in terms of the roles that each plays in the species that display them.

The evolutionist's line

With that little propadeutic exercise out of the way, let me get to the main arguments I want to make in this chapter. The proposition to be defended is that the study of unconscious processes generally and implicit learning specifically should be cast into an evolutionary setting. I will try to show how such a perspective provides a different but provocative explanatory framework for implicit learning and yields some novel and intriguing insights into unconscious cognitive processes. We will allow evolutionary biology to act as an explanatory vehicle for understanding implicit, unconscious mentation and for differentiating these covert processes from explicit, conscious thought.

There are two things I will not attempt to do here. First, I will not deal with the various "information processing" models that have been put forward to handle some of the experimental findings of implicit learning. Whatever I have to say about the phylogenetic history of primitive induction routines can be evaluated and explored independent of any formal instantiation of the processes involved. The success or failure of a PDP machine (Cleeremans, Servan-Schreiber, & McClelland, 1989, in press; Cleeremans & McClelland, 1991) to carry out an induction of the underlying structure of an artificial grammar will have little bearing on the arguments I want to pursue. Second, I will not attempt to deal with the flip side of this issue, the evolution of consciousness and how it emerged from a perceptual/cognitive system devoid of awareness. As I argue in the next chapter, this is a complex process with at least two critical components. The first involves explication of how primitive organisms came to recognize that there was "something out there," that is, the recognition of externality. The second involves the deeper question of the emergence of an awareness of self with the capacity for self-referencing. Together they represent a class of problems that requires more bravery (or perhaps foolhardiness) than I have; I leave the task to such intrepid scholars as Jaynes (1976) and Dennett (1991).

The largely speculative nature of this endeavor is, as I hinted above, due in large measure to the simple fact that there has been, to date, rather little direct empirical evidence put forward in support of the assertions made. However, as will become plain, the orientation taken derives from standard heuristics that are as much an aspect of evolutionary biology as the principle of natural selection itself. The model developed, indeed, rests on a firm foundation, and the data that do exist are quite in keeping with it.

This evolutionary framework also happens to be one within which I have always felt comfortable. It has proven fruitful in that it makes fairly "clean" predictions that are testable in a straightforward manner. It also, of course, goes hand in glove with the functionalist/adaptationist stance I have adopted toward cognitive function. Moreover, as I will argue in more detail in the next chapter, the overall metaphysics of the functionalist's approach is more moderate than either of the two primary competitors, the strong nativism of the Chomskyan or Fodorian kind on one hand and the radical behaviorism of the Skinnerian type on the other. Finally, from a methodological point of view, it is also broadly eclectic in its procedures and provides, I believe, the kind of balanced focus that is required if we are to make sense out of the complexities of the human condition.

Turning now to the details, we need to introduce several basic principles of evolutionary biology that will form the foundation for the arguments to be developed.

An evolutionary context for the cognitive unconscious

Some background

Evolutionary considerations have been entertained by several of those working on implicit processes. The first systematic analysis was offered by Rozin (1976), who developed a Darwinian context for the evolution of intelligent behavior. Rozin argued that intelligence, treated as a phenotype, should be viewed as being organized in a hierarchical manner out of older, primitive, and encapsulated modules, each of which had evolved restricted programs for particular functions. The key to Rozin's model is the assumption that over evolutionary time the modular units become more and more accessible to each other, resulting ultimately in the emergence of a consciousness capable of a broad range of executive control over various behavioral and mental functions. In Rozin's conception, the individual, hard-wired modules are assumed to function independently of consciousness—indeed, taking the current perspective, they can be seen as the underlying mechanisms for implicit cognitive processes.

The modular notion is also reflected in the phylogenetic proposals of Schacter (1984) and Sherry and Schacter (1987). Schacter has consistently argued that human memory needs to be viewed not as a unitary process but as one made up of a (still undetermined) number of separate systems. Sherry and Schacter (1987) explored the feasibility of various evolutionary processes that might have given rise to a memorial system composed of multiple components with dissociable functions. Their proposals differ from Rozin's primarily in that they give a greater role to the process of exaptation (Gould & Vrba, 1982), whereby existing systems that evolved for one function might assume various other functions (also see Gould & Lewontin, 1979). Both models, however, are significant in that they emphasize *function* rather than being satisfied with descriptions and analyses of mechanisms.

While these two are the most thoroughly developed evolutionary analyses of the cognitive unconscious, other researchers have made provocatively similar suggestions. Squire (1986), in an overview of the mechanisms of memory, noted that procedural memory appears to be subserved by neurological systems that are phylogenetically older than those that underlie declarative memory. He pointed out that primitive associative processes, which represent the first form of a procedural system, are handled by neuronal systems, which occur in invertebrates, while declarative systems, which involve consciousness and awareness, require the elaboration of the medial temporal structures, including the hippocampus. A similar proposal was made by Shimamura who suggested, almost as an afterthought (see Shimamura, 1986, p. 108), the equally intriguing proposition that the systems that subserve procedural knowledge might also develop ontogenetically earlier.

There are similarities between these evolutionary approaches and the one taken here. There are also differences. The emphasis on function and the adaptive value of structure and form is carried over but not the modular-systems perspective. It is not that I feel that models based on multiple systems are invalid, it is merely that I do not see how, at this point, to either accept or reject them. The current data base seems to be roughly amenable to either interpretation, and with a modicum of ingenuity, equally good "just-so stories" can be constructed to provide an adaptationist picture for either scenario.

Moreover, no absolute dichotomy is necessary in order for the following arguments to go through. The dissociations drawn between the implicit and the explicit are predicated primarily on the identification of factors and variables that display differential effects on functions that have manifestly implicit or explicit components. Such factors and variables, of course, will display their effects without regard to whether the underlying mechanisms are multiple or unitary in nature. Moreover, I have a general suspicion of our long-standing tendency for Balkanizing the mind; we tend to hypothesize "new" types of learning and memory on insufficient evidence. Given less-than-compelling arguments on behalf of a multiple systems model, I prefer to err on the conservative side.

Some heuristics of evolutionary biology

There are several, basic considerations that have dominated evolutionary biology for over a century and a half. Many of these heuristics have been so successfully absorbed into the field that they are rarely explicitly stated any more. However, they have an important history, and it will help in the development of the arguments below if we review here (1) von Baer's early laws of embryological development, and (2) Wimsatt's contemporary *developmental-lock* model of evolutionary change.

VON BAER'S LAWS. In 1828, Karl Ernst von Baer, the German embryologist, put forward four laws of development. Von Baer's formulations were articulated as part of his extended critique of the Haekelian doctrine that ontogeny recapitulated phylogeny, a doctrine that dominated (as any position that comes to be called a "doctrine" must have) embryological thought then and for decades to follow (see Gould, 1977b). The essence of von Baer's arguments was that the embryos of higher species did not and could not possibly pass through the adult forms of lower species as the recapitulationists maintained. The four laws (using Gould's 1977b translation) express this rather succinctly:

1. The general features of a large group of animals appear earlier in the embryo than the special features.

2. Less general characters are developed from the most general, and so forth, until finally the most specialized appear.
3. Each embryo of a given species, instead of passing through the stages of other animals, departs more and more from them.
4. Fundamentally therefore, the embryo of a higher animal is never like the adult of a lower animal but only like its embryo.

These laws are now captured through the generalization that *differentiation proceeds from the general to the particular,* a principle that became, after a period of unfortunate neglect, essentially axiomatic in evolutionary biology.[3] It is important to note that von Baer articulated his laws in this fashion in order to deal with questions of embryology and ontogenesis. They were tailored to show the impoverishment of the recapitulationists' position. However, they also provided a framework within which to see greater depth in the relationship between ontogeny and phylogeny. That is, the recapitulation is not "literal" and not with adult forms, as Haeckel had argued. Rather, it is the case that one can identify deep principles that govern development at both the ontogenetic and the phylogenetic levels.

WIMSATT'S PRINCIPLES. In a recent series of papers, Wimsatt and his co-workers have pointed out (Schank & Wimsatt, 1987; Wimsatt, 1986, in press) that von Baer's pre-Darwinian formulations, as well as a variety of other basic principles of contemporary evolutionary biology, can be given a formalization that they have called the *developmental-lock model.* The notion of a developmental lock emerges from considerations of the hierarchical manner in which complex systems are constructed or evolve. The formalization derives, as noted above, from the earlier considerations of the nature of complex systems presented in the classic paper on such systems by Simon in 1962.

The easiest way to capture Wimsatt's arguments and to understand why he refers to it as a *developmental*-lock model is to use the metaphor of a cylindrical combination lock, just like the common bicycle lock that has some number of wheels each of which must be set to its proper position for the lock to open. The mechanism that controls such locks is a set of wheels, each of which has some number of positions it can be turned to, only one of which is "slotted," allowing the anchoring shaft to slide in. A sample lock is shown in Figure 3.1.

There are at least three interesting kinds of locks one could construct using this basic mechanism. Wimsatt refers to them as *complex, simple,* and *developmental.* First, one could have a *complex* lock in which someone trying to open the lock has no clue as to the correct position for each of the *n* wheels. Here the number of possible combinations is very large, depending on the number of

[3] This despite the fact that von Baer was never an evolutionist and certainly never persuaded of the mechanisms of Darwinian natural selection. Throughout a long career that extended well into the years of Darwin's ascendance, he remained a firm teleologist and a believer in ideal progress toward ideal forms.

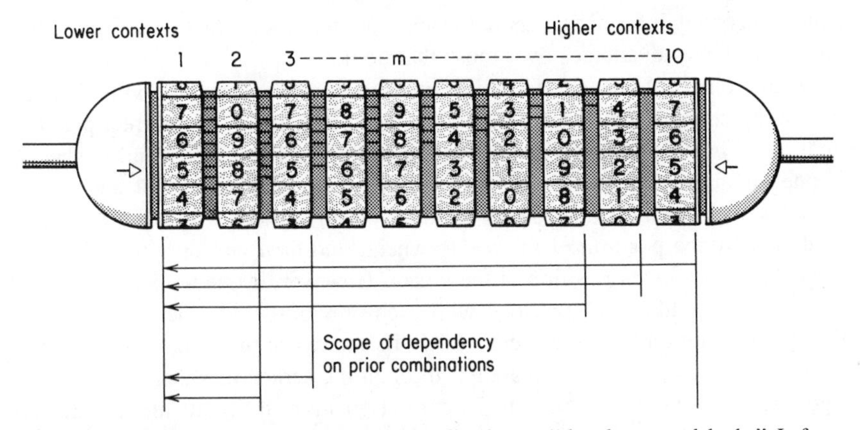

Fig. 3.1. A typical cylinder lock conceptualized as a "developmental lock." It functions as a "complex" lock if worked from right-to-left and as a "simple" lock when worked from left-to-right. In the left-to-right form it captures the hierarchical principles of development with the later emerging "higher contexts" predicated on the prior combinations established for the "lower contexts." Any change in position of any wheel randomly resets the combinations for all wheels to the right of it. This figure is adapted from "Generative entrenchment and evolution" by J. C. Schank and W. C. Wimsatt, 1987, *PSA 1986, 7,* p. 35. Copyright 1987 by the Philosophy of Science Association.

positions on each wheel and the number of wheels, and the likelihood of solving[4] the lock very small. If there are n wheels and m positions on each the expected number of trials is $n^m/2$. Of course, this is the way such locks are ideally constructed in the real world.

One could also have a *simple lock,* in which setting each wheel at its correct position is accompanied by a subtle clue, like a faint click. In this case, the number of combinations someone attempting to open it would need to try is dramatically lessened and given by $m/2 \times n$. (This, of course, is why expert safecrackers use stethoscopes and other sensing devices.) Note, importantly, that both the simple and the complex locks are *directionally independent;* they can be opened equally effectively by moving from either left-to-right or from right-to-left. The computational advantage that the simple lock enjoys over the complex was taken by Simon (1962) as a general heuristic expressing the advantages of reductionistic or analytic approaches to the analysis of systems where the complex structures are decomposed in elementary parts. That is, complex sys-

[4] The term "solution" here is used metaphorically to stand for the establishment of either a viable species (phylogenetic solution) or an individual organism (ontogenetic solution). The general principles of the Schank and Wimsatt developmental lock model operate in both cases. Wimsatt's research program is much larger than the piece I have selected here. It is designed to provide a biologically and philosophically coherent, formal framework within which to model the notion of "generative entrenchment," or the extent to which particular forms become established in evolutionary development. For details, see Schank and Wimsatt (1987) and Wimsatt (1986; in press).

tems that convey information about the appropriateness or effectiveness of earlier operations have advantages over those that do not.

Finally, one could have a *developmental lock* in which the solution is dictated by the direction in which one attempts to open it. That is, the correct position on the nth wheel is governed by the position of the $n-1^{st}$ wheel, for all n wheels. If one attempts to open the lock by working from right to left, it acts like a complex lock because the correct position on any wheel will be undermined randomly by the position of the $n-1^{st}$ wheel, and there are no clues as to the correctness of a given position of the wheel. However, if one works it from left to right, it acts like a simple lock with each previously set wheel determining the correct solution for the current wheel. To appreciate this, think of the solver attempting to pull the anchoring shaft through the series of wheels from left to right. As each wheel is turned to its correct position, the shaft may be moved one position (which is why real cylinder locks have a safety catch that prevents the shaft from being moved until all wheels are in their correct position).

The developmental lock, of course, is the interesting metaphor from the point of view of evolutionary biology. Indeed, in its left-to-right solution form with the earliest stages of development on the left and later stages further to the right, it functions as a powerful model for both phylogenetic and ontogenetic analyses. Metaphorically, each wheel is a developmental feature (either ontogenetically or phylogenetically). The time line moves from left to right with the leftmost wheels conceptualized as features of greatest antiquity. Schank and Wimsatt (1987) show that a number of important general principles of evolutionary biology are proper theorems of the formalization of the developmental-lock model—including von Baer's laws. The following seven theorems from the 1987 paper are relevant to the thesis that I am developing here.[5]

1. Evolution is increasingly conservative at earlier stages of development; features that emerge at earlier developmental stages change at lower rates.
2. Most adaptive evolution takes place through modifications that occur late in development.
3. Features expressed early in development have a higher probability of being required for those expressed later.
4. Differentiation proceeds from the general to the particular.
5. In general, features expressed earlier in development are evolutionarily older.
6. Features that have persisted for a significant time tend to become ever more resistant to change as more and more features become dependent on them.
7. Comparative, cross-species analyses can generate significant information about the structure of developmental programs.

[5] The formalization of the developmental-lock model allows for an extended set of theorems beyond those presented here. For formal proofs of these and the other propositions that make up parts of the general model of generative entrenchment, see Schank and Wimsatt (1987).

Some generalizations

It is possible to distill the above propositions of von Baer and Wimsatt down to four, much more general principles, as follows:

1. *The Principle of Success.* Once successful forms emerge, they become the foundation for later forms. All of evolution is testimony to this principle. It also is a simple entailment of the developmental lock model.
2. *The Principle of Conservation.* Following from the above, developmental processes are conservative. Once successful forms are established, they tend to become fixed and serve as foundations for emerging forms. For example, once the set of jaw bones in early fish became modified and began to function to transmit vibratory information, the form of the inner ear was fixed. Virtually every species that has an auditory sense uses these same three ancient bones in this general fashion. This principle follows directly from both von Baer's first two laws and Schank and Wimsatt's first, second, third, and fourth theorems.
3. *The Principle of Stability.* Following from the preceding, earlier appearing, successful, and well-maintained forms and structures will tend toward stability, showing fewer successful variations than later appearing forms. Under some circumstances, variations are selected for and under other conditions they are not. Given that the environment remains stable, it becomes relatively unlikely that variations that occur in earlier forms will prove adaptive. Hence, the relatively fixed, successful forms will display greater stability than the later appearing forms. The stability principle is reflected by the cladistic classification systems in evolutionary biology, which are based on there having existed an ancestral species from which the members of the clade are descended. This principle is embodied in Schank and Wimsatt's fifth and sixth theorems.
4. *The Principle of Commonality.* Evolutionarily earlier forms and functions will be displayed across species. This principle, like the others, follows naturally from the preceding. Von Baer's fourth law generalized to the phylogenetic level clearly entails it, and it is directly embodied in Schank and Wimsatt's seventh theorem. It also lies at the heart of much of ethology and comparative psychology and, of course, is a virtual axiom of behaviorism.

These general principles should be viewed as heuristics of evolutionary biology. That is, they are not laws in any strict sense and should not be seen as inviolable. They are more like "operating principles," and generally speaking, their value derives from their capacity to guide research and theory. In order to link these principles up with the material reviewed in the first two (or three) chapters and put the cognitive unconscious into an evolutionist context, we only need one more step: A simple axiom.

An Axiom About Consciousness

> Consciousness is a late arrival on the evolutionary scene. Sophisticated unconscious perceptual and cognitive functions preceded its emergence by a considerable margin.

Before developing the various hypotheses about implicit and explicit cognitive functions that can be derived from this assumption, two short asides are needed to clarify the nature of the argument being developed.

AN ASIDE ON THE MEANING OF "CONSCIOUSNESS". I meant the axiom to be straightforward and noncontroversial, and as a statement about evolution, it is provided so we can achieve some clarity on key terms. I recognize that up to now I have been using the terms *conscious* and *consciousness* in a fashion that is, to be charitable to myself, "fast and loose." Although my culpability here is shared with most of my peers, this is as a good a place as any to make some attempts to be clear about what is covered by these terms and exactly what kind of *consciousness* is being referred to in the axiom. Although I am not prepared for a full analysis of what is meant by the term *consciousness,* there are some aspects to it that do need to be understood. Later, in Chapter 4, I will make another foray into this topic.

I want to distinguish between a consciousness that consists essentially of an "awareness" of an external world that is not oneself and a consciousness that carries an "awareness" of and the capacity to play a causal role in the inner workings of oneself. There seems little doubt that the former variety of mentation encompassing an awareness of the environment, of self, and of others, existed and still exists in primitive species. There is similarly little doubt that the latter, represented as a set of executive functions with causal roles to play in mental life, finds a unique expression in our species. Weiskrantz (1985) suggested that the distinction here may be captured by the fact that the latter variety involves the conscious process of *monitoring*.

The variety of consciousness of concern here is one I view as an emergent aspect of a complex brain, and although various forms of awareness of the environment, of self, and of others likely existed and still exist in various primitive forms in lower organisms, consciousness achieves a unique executive function in *Homo sapiens.*[6] We can prevent a good deal of confusion if we are careful to

[6] Occasionally such proposals are viewed as running counter to the standard orientation in contemporary cognitive psychology, which is strongly physicalistic in nature. I do not feel that there is any inherent contradiction here. Consciousness, in my view, should be treated as a kind of biological metaphor, a "psychological" abstraction for an underlying complex of physiological and neurological actions, not unlike the way in which we have long treated similar abstractions like *mind, image,* or even *ego.* The question of whether it makes sense to construct theories based on the presumed "reality" of mental events or whether this exercise is simply "something to do till the physiologists get here" is really a question of pragmatics. I have no trouble presuming that mental states can be theoretically conceptualized as having causal roles to play in behavior while simultaneously accepting the (reductionistic) argument that these

distinguish this uniquely human version of consciousness from the kinds of awareness of the external world that it would seem most reasonably sophisticated locomotive species possess (see Griffin, 1981, 1984). In some ways it is unfortunate that we don't have separate terms for these different mental functions. Piaget called the kind of conscious functioning with which I am concerned "cognizing" to differentiate it from the simpler forms of awareness that are essentially reactive and passive, but unfortunately, the term never caught on.

Hence, the axiom is quite specifically meant to encompass this latter variety of consciousness, the one we link with the capacity for self-referencing and for characterizing and explicating the functions and actions of self. Were it otherwise, the model I am developing would have little to say of interest on the topic of implicit cognition.

AN ASIDE ON PRIORITIES. It's worth pointing out here that the form of the argument I am putting forward should have the effect of "flipping," as it were, the priorities of unconscious and conscious cognition. Traditionally the focus in psychology has been on consciousness with the implication that unconscious processes should be dealt with by exclusion; only if you failed to show that a process was conscious could you conclude that it was unconscious. A look at the history of the many ways in which the unconscious has been handled shows this bias clearly (Ellenberger, 1970; Erdelyi, 1985). Consciousness tends to assume epistemic priority largely because it is so introspectively obvious; unconscious systems are thus left to be dealt with derivatively. As Kihlstrom (1990) has pointed out, psychology's very beginnings as a distinct field of scientific exploration began with the examination of consciousness as its defining characteristic. As I argued in Chapter 2, this perspective has its roots in our Lockean foundations.

The position that consciousness is primary has forced a distinctly skewed view of unconscious cognitive functions with some serious attendant problems. One such difficulty has emerged recently in various disputes over whether or not particular kinds of mental functions can "really" be regarded as unconscious (see here Holender, 1986, and various commentaries on that paper as well as the points of view put forward by Dulany, Carlson, & Dewey, 1984, 1985; Brody, 1989; and Perruchet & Pacteau, 1990). The debate here has had unfortunate consequences because it has been reduced to squabbling over whether a particular methodology has or has not completely removed the possibility that some flickering residue of awareness remains in the subject's mind. As Erdelyi (1986) pointed out, and as I argued in Chapter 2, no such conditions can be achieved in any practical laboratory setting and, in any event, such a requirement is ultimately irrelevant. The only critical demonstration is that there exists some unconscious knowledge base that exceeds the conscious (see Reber, 1989b).

---

mental states could, in principle, be represented in neurophysiological form. The point, of course, is that one will be able to gain more *insight* into function by theorizing about mental states than by theorizing about physiological functions—even if they were well understood.

The proper stance, of course, which the evolutionary orientation makes plain, is the one that emphasizes the primacy of the unconscious (Reber, 1990). Such a perspective restores to the unconscious the empirical and philosophical priority it deserves but has not enjoyed since before the British Empiricists. The unconscious mental processes are the epistemic foundations upon which emerging conscious operations are laid. Such a position also rescues us from the "methodological sophistry" that the other ordering invites (see Reber, 1989b). It is heartening that several other theorists have come to support this reordering of priorities, specifically, Rozin (1976), Kihlstrom (1987, 1990), and Lewicki (1986a; Lewicki & Hill, 1989).

To summarize, unconscious, implicit, covert functions must have antedated conscious functions by a considerable period of time. Therefore, invoking the standard heuristics from evolutionary biology, we should anticipate finding that implicit cognitive processes generally, and implicit learning specifically, display a variety of properties that differentiates them from the overt, explicit, and conscious forms of mental function. Specifically:

Hypothesized characteristics of implicit systems

1. *Robustness:* Implicit learning and implicit memories should be robust in the face of disorders and dysfunctions that compromise explicit learning and explicit memory.
2. *Age independence:* As compared with explicit learning, implicit acquisition processes should show few effects of age and developmental level.
3. *Low variability:* The capacity to acquire knowledge implicitly should show little in the way of individual-to-individual variation. Population variances should be much smaller when implicit processes are measured than when explicit processes are.
4. *IQ independence:* Unlike explicit processes, implicit tasks should show little concordance with measures of "intelligence" assessed by standard psychometric instruments such as the commonly used IQ tests.
5. *Commonality of process:* The underlying processes of implicit learning should show cross-species commonality.

Each of these needs a little developing.

Robustness

If implicit learning and the tacit knowledge base that results from it are based on evolutionarily old structures and processes, then they should show greater stability and resiliency than acquisition processes and their memorial products that are founded on more recently evolved structures and processes. The unconscious should be more robust than the conscious. Of the five hypothesized principles, this is the one for which there is the largest and most unambiguously supportive data base.

IMPLICIT MEMORY. The implicit process most often studied here is implicit memory. That is, memory for information once acquired, which can be shown to effect behavior but which is not available for conscious retrieval. Some of this work was outlined in the historical overview in Chapter 1. Here, I select the findings that most compellingly suggest the evolutionary model. Implicit memories can be created in normal subjects in laboratory settings by presenting material so that it remains outside of awareness. Standard techniques here include the use of rapid, masked, visual presentation of stimuli (Kunst-Wilson & Zajonc, 1980; Marcel, 1983a), dual tasks in which the critical information is presented in the unattended channel (Corteen & Wood, 1972), and parafoveal visual presentation in which the stimuli are presented in the periphery of the visual field where acuity is poor (Bradshaw, 1974).

Although there is considerable controversy over some of these findings (see Holender, 1986, and the various commentaries on it), it is clear that memories can be established that resist attempts at conscious retrieval but, nevertheless, display their effects on later behavior. In a recent review of the literature, Greenwald (1992) concludes that most of the paradigms that have been used to explore unconscious perceptual and cognitive functioning have yielded supportive data. For instance, Kunst-Wilson and Zajonc's (1980) now classic finding of affect-based discrimination in the absence of conscious recognition has been replicated and extended by Seamon and his co-workers (Seamon, Brody, & Kauff, 1983; Seamon, Marsh, & Brody, 1984). In these studies, subjects were shown stimuli under such degraded conditions that they functioned at chance levels when asked to select which of two stimuli they had seen. Yet they showed evidence of having picked up information about the presented stimuli by selecting a previously presented figure when asked which of two figures they "preferred." Implicit information encoded in an affective form has an impact on choice behavior independent of any conscious apprehension of that information. Assuming that the processing system(s?) that modulates the affective form of encoding is phylogenetically older and more primitive than the one(s?) controlling conscious recognition (which seems like a safe presumption), such a result fits within the evolutionary thesis.

While these findings are important in supporting the argument that implicit and explicit memories are distinct, they speak only indirectly to the evolutionary hypotheses entertained here. The more interesting cases are those in which memories have been rendered nonretrievable by virtue of neurological insult or injury, in which the absence of awareness is a direct result of tissue damage and not an experimenter's sleight of hand. Many of the methodological criticisms levelled by Holender (1986) against the work just reviewed are rendered moot by examining these subject populations.

There is a large clinical literature showing that implicit memories remain robust in the face of dysfunctions that impair retrieval (see Schacter, 1987; Schacter, McAndrews, & Moscovitch, 1988; and Shimamura, 1989, for excellent overviews). Graf, Schacter, and their colleagues have showed that amnesiacs who have essentially no recognition of stimuli previously presented and no capacity for their conscious recall still show implicit memory for these stimuli (Graf & Schacter, 1985; Graf, Squire, & Mandler, 1984; Schacter, 1987). Nissen, Knopman, and Schacter (1987) also presented evidence that pharmacologically induced memory failure for previously presented stimuli (brought about by administration of scopolamine) left implicit memory for a repeating sequence of events intact. Shimamura (1986) recently reviewed dozens of experimental reports showing that amnesic patients exhibit intact priming effects even though they have little or no capacity for recall of the previously presented information.

In a typical priming study, Jacoby and Witherspoon (1982) presented a group of Korsakoff patients and a group of normal control subjects with a series of simple tasks, like "name a musical instrument that employs a reed." They later asked all subjects to spell homophones such as *reed/read* and found that both groups showed a distinct bias for the previously presented form. The normal subjects, of course, could recall the previously presented questions, the Korsakoff amnesics could not.

This literature is generally regarded in neuropsychology as providing evidence for dissociable memory functions. The suggestion has been put forward, specifically by Squire (1986) and Shimamura (1986), that this dissociation may be viewed in evolutionary terms. Shimamura has linked the implicit memory system with procedural memory and argued that, as such, it is handled independently of the hippocampus and its associated input/output connections. These hippocampal functions are generally considered to be part of a phylogenetically (and perhaps even ontogenetically) earlier memory system than explicit, retrieval processes. The neurological insult that produces the amnesia can, therefore, be seen as having a greater effect on the more recently evolved and more fragile process, the complex retrieval mechanism associated with hippocampal function, than it has upon the older, more robust process of information storage (whose neurological sites are still unknown).

While all this is rather compelling, there are additional etiological and neurological issues that may prove to be of importance. For example, the amnesic syndrome is not a singular disorder. Amnesic patients may be display either a retrograde or anterograde form or, on occasion, they may show signs of both dysfunctions. The disorder may be due to any of a wide variety of precipitating events including surgery, alcoholic Korsakoff's syndrome, Huntington's disease, ischemic episodes, hypoxic episodes, encephalitis (of varying forms), traumatic head injury, closed head injury, vascular disorders (of varying kinds), and tumors, and of course, it may be accompanied by various other disabilities, as in Korsakoff's, Huntington's, and Alzheimer's cases, where a variety of other more global impairments are typically observed.

Moreover, amnesias have been reported in cases where the site and extensiveness of the tissue damage ranges considerably. For example, there are documented cases of neuropathology ranging from the hippocampal areas (as in the famous case of H. M.), bilateral lesions of the walls of the third and fourth ventricles (as in Korsakoff's syndrome; see Shimamura, Jernigan, & Squire, 1988), widespread neuropathology of various cortical structures including the orbitofrontal cortex and cingulate areas as well as bilateral damage to the basal forebrain (as in a case of post-viral encephalitis amnesia; see Damasio, Eslinger, Damasio, Van Hoesen, & Cornell, 1985), and any number of more localized lesions produced by specific insult or injury including an unusual case of a fencing injury in which a miniature foil entered the patient's right nostril and penetrated the left diencephalon and caused extensive midline damage to thalamic nuclei and mammillary nuclei (Squire, Amaral, Zola-Morgan, Kritchevsky, & Press, 1987; Teuber, Milner, & Vaughn, 1968).

The hippocampal regions are certainly the ones most often assumed to be critical to memorial functions (see Rawlins, 1985), and they are also frequently implicated in the amnesias. In addition, as Gray (1982, 1984) points out, the hippocampus also appears to have important monitoring functions. But it would be a mistake to assume that somehow conscious monitoring functions of explicit memory are "localized" in these regions. The current picture is a confusing one because the syndromes that have been implicated in the implicit/explicit dissociation are so varied and the neuropathologies so diverse.

The point is that here, like in other problem areas dependent on classification schemas, it is not always obvious which syndromes and disorders belong together and which need to be kept distinct. Recent work by Butters and his coworkers helps to make this point. After a review of the literature on the memory disorders found in Korsakoff's, Alzheimer's, and Huntington's patients, Butters, Salmon, Heindel, and Granholm (1988) concluded that these disorders can be distinguished in a reasonably clear manner. Huntington's and Alzheimer's patients both show the classic amnesia for explicit memories but for different reasons. Huntington's cases have intact capacity to store new information but suffer from seriously impaired abilities to initiate systematic retrieval processes. Alzheimer's patients have difficulties in storing new information explicitly although they have some intact implicit memories.

Moreover, when motor skills are examined, additional dissociations are implicated. Huntington's patients are typically severely impaired in pursuit rotor tasks (Heindel, Butters, & Salmon, 1988) although they show near normal performance on stem-completion priming tasks. The opposite is found with the Alzheimer's patients, who are near normal on motor skills but show severely compromised performance on the priming tasks (Moss, Albert, Butters, & Payne, 1986). Butters (1989) compared Huntington's and Alzheimer's patients and normal control subjects on five distinct implicit tasks: stem completion, semantic priming, perceptual priming, pursuit rotor, and weight biasing (in which memories for earlier hefted weights are displayed by biases in later weight-estimation tasks). Alzheimer's patients showed little lexical, semantic, or pictorial priming but were indistinguishable from normals on the pursuit-rotor and weight-biasing tasks. Huntington's patients were essentially the mirror image, with virtually normal priming in all three stimulus types, no learning of the pursuit rotor task, and a partial effect on weight biasing.

Butters's assessment of these findings is that the implicit tasks may not be quite so monolithic as some have supposed, that the locus of the neuropathology may turn out to have a significant role in the degree to which particular functions survive. However, an intriguing aspect of this pattern of findings is that they follow phylogenetic lines. Alzheimer's patients have extensive cortical and subcortical disturbances that severely compromise virtually all standard cognitive functions; the remaining "cognitive" functions are those that subserve the evolutionarily older sensorimotor tasks. In the Huntington's patients, the specific neuropathology in the basal ganglia disrupts motor functions but, being largely outside of the cognitive domain, does not compromise the priming tasks.

In summary, the extensive literature on implicit and explicit cognitive functions in various special populations with neurological, psychological, and pharmacological disorders supports the general dissociation of function into two forms, one that is implicit, nonreflective, and procedural, in short, *unconscious,* and one that is explicit, reflective, and declarative, in short, *conscious.* These two categories of function are also dissociable along phylogenetic lines, with the evolutionarily older, implicit system showing the greatest resistance to insult and injury. In cases that do not fit neatly within the implicit/explicit classification, like the results from Butters's studies, they appear to follow the predictions of the phylogenetic model.

IMPLICIT LEARNING. Note, however, that these studies all examined memorial or operating systems concerned with the temporary storage and retrieval of material already known by the subjects. As Rozin (1976) put it, these implicit memory effects all reflect the activation of *preexisting representations.* For example, all of the subjects in Jacoby and Witherspoon's study knew the words "reed" and "read" and how to spell them when they entered the experimental setting. Butters's (1989) subjects all knew the various words and morphemes used in the stem-completion, semantic-priming, and perceptual-priming experiments, and while it is likely the case that the motor skills his group has examined were new to his subject populations, the issue of motor learning is tangential to the focus of this inquiry. This issue of preexisting knowledge is an important point that is often glossed in work in this area because relatively few of the studies with dysfunctional subjects actually look at "real" learning, the acquisition of information in the experimental setting and not merely the recombing or reactivation of information brought to the experimental setting.

True, implicit learning has been examined only recently in these specialized populations. In the previous chapter, various conditions were outlined for the proper investigation of implicit learning; one of these is that the material to be learned needs to be novel and not part of the subject's "preexisting representations" when they begin the experiment. That is, the kinds of learning of interest here are those that involve the acquisition of something like a set of rules for symbol ordering (as in the studies with artificial grammars) and not the learning of an arbitrary association between two familiar words (as in many of the implicit memory studies).

I know of only five studies, all carried out recently, that have used these kinds of stimulus materials in studies with dysfunctional patient populations. The findings all support the general picture being developed here. In one study, Abrams and Reber (1988) found that a mixed group of psychotic patients showed a capacity to use implicitly acquired knowledge about an artificial grammar that was not distinguishable from that of a group of college students. This experiment was a standard AG learning study in which both normal and psychotic subjects first had to memorize exemplars from the grammar and then to distinguish grammatical from nongrammatical strings in a well-formedness test. Interestingly, in this study the psychotic patients had more difficulty with the memorization phase of the experiment than the controls, taking more trials to reach criterion; on the well-formedness task, which is, as argued in Chapter 2, the true test of implicit functioning, their performance was indistinguishable from the normals. However, in a separate experiment in which the task was changed to one on explicit problem solving, psychiatric inpatients showed dramatically poorer performance than the normal controls.

Knowlton, Ramus, and Squire (1992) recently extended these findings in a study of amnesic patients. They reported that amnesics showed fully intact artificial-grammar learning when run using the standard procedure. However, when Knowlton et al. made a simple adjustment in the instructions to the subjects prior to running the well-formedness task, amnesics performed well below the control subjects. This simple modification was to instruct all subjects to attempt the well-formedness task by recalling the specific strings used during learning and using them as the basis for decision making. Under these conditions the amnesics were no longer able to perform the task. The functional representations established during the memorization phase were only available for decision making when a nonconscious mode was operative.

In a similar vein, Knopman and Nissen (1987) demonstrated implicit learning of a repeating pattern in (probable) Alzheimer's patients who were generally incapable of acquiring such complex information under the standard overt learning conditions. Also here, Johnson, Kim, and Risse (1985) presented Korsakoff patients and normals with repeated melodies from traditional Korean songs. After a short delay, subjects were asked to give preference ratings of a series of melodies, some of which were the ones they had been presented with earlier. Both groups showed the same preference bias for the repeated melodies, but only the normals showed recognition memory for them.

Finally here, Squire and Frambach (1990) used Berry and Broadbent's (1984) production control tasks with a group of amnesic patients. First, amnesics and normal controls interacted with the impersonal, sugar production task after which they were given a questionnaire that examined their explicit knowledge of simple facts about the procedure as well as both specific and general strategies that they may have used. This procedure was followed by a similar opportunity to work with the person-interaction task. All subjects were brought back to the lab 27 days later and run through the full experiment a second time.

On the initial run through the two tasks, there were no differences between groups, suggesting that both groups were using the implicit processing system.

However, after the 27-day interval the control subjects were superior to the amnesics, which Squire and Frambach argued was likely due to the normals gradually building up an explicit knowledge base that could be used to improve performance. The questionnaire data generally supported this interpretation in that the amnesics were significantly poorer in their knowledge of simple facts about the experiment and marginally poorer in their knowledge of general strategies.

These five experiments are important in that they address questions that have been raised (see Shimamura, 1989) concerning whether or not it is possible for such implicit acquisition processes to remain following the kinds of cognitively devastating insults that accompany such disorders as Korsakoff's syndrome and Alzheimer's disease. These experiments suggest that such learning is indeed possible when the proper acquisition procedures are implemented.

To summarize, a rather broad expanse of research on problems as diverse as affective discrimination of obscure geometric forms that cannot be recognized to implicit learning of complex artificial grammars in psychotic and amnesic patients can be seen as speaking to the same deep cognitive process. Taken as a whole, the full collection of studies here provides strong support for the general proposition that implicit systems are more robust than explicit systems. The evolutionary stance provides a coherent theoretical package for these findings and, interestingly, suggests that particular neurological loci should eventually be discovered to be playing important roles in these two types of processing. I will resist speculating about what these structures might be other than to suggest that evolutionary antiquity and phylogenetic considerations will likely prove to be important aspects of future models.

Age Independence

The simplest argument in support of this principle is the most obvious: children must be capable of implicit acquisition of complex knowledge of their environments because that is what they do. Children acquire stunning amounts of information about their physical, social, cultural, and linguistic environments at a very early age and do so relatively independently of conscious attempts to acquire that information and without much in the way of conscious knowledge of what they have, in fact, learned.

However, when it comes to hard data for this and the remaining principles discussed below, there is nothing like the empirical support that we found for the robustness principle. Generally speaking, implicit cognitive processes have not been the focus of systematic research programs in developmental psychology; the work in the field is typically carried out without regard to the kinds of issues that the theory of implicit learning has raised. I am aware of only a few research programs that give what could be viewed as "semidirect" evidence for this hypothesized principle.

First, there is the literature on the related problem of automaticity where procedural knowledge is displayed largely outside of consciousness. Hasher and Zacks and their colleagues have argued that such highly automatized functions as keeping track of the frequency of events show little in the way of age effects

(Hasher & Chromiak, 1977; Hasher & Zacks, 1984). Their position has an evolutionary component in that they feel that highly automatized functions such as encoding the frequency of events or noting the locations of events in the environment must have played a critical role in the survival of primitive organisms that possessed them. However, there are reasons for not generalizing too strongly from the typical Hasher and Zacks tasks to the tasks typically studied in experiments on complex implicit learning. One of the features that characterizes the automatized processes studied by Hasher and Zacks is that they require little in the way of attentional resources. There are many reasons for suspecting that true implicit learning of the underlying nature of complex stimulus displays requires considerable attentional focus (see Dienes, Broadbent, & Berry, 1991; Sanderson, 1989, 1990). However, even though implicit learning may turn out not to be as "automatized" a processes as the logging of frequency and location characteristics studied by Hasher and Zacks, there are good reasons for suspecting that both processes, by virtue of their "implicitness," will show little ontogenetic variation.

Second, Brainerd and his colleagues have recently developed an approach to cognitive development that argues that the role of intuition has been underplayed in contemporary theoretical work (see Brainerd & Reyna, 1990). This conception of knowledge acquisition emphasizes the role of general processing systems that operate by extracting the gist from particular settings, engaging in inexact development of preferences and parallel processing of inputs, among other procedures. Brainerd's approach is much closer to theories that stress heuristics, intuitionism, and inexact thinking than to those that emphasize either developmental stages such as Piaget's or the information process metaphor found in many contemporary theories in cognitive science. While there are many differences between Brainerd's work and the existing literature on implicit learning, the important point is that the sharp age effects that are the hallmark of much of contemporary developmental work are not so prominent when these "fuzzy" (Brainerd's term) cognitive systems are engaged.

Third, there is the developmental literature on implicit memory. The results are generally in agreement with the age-independence hypothesis, although the picture is a complex one. The literature includes quite a few examinations of memory in infants and young children. In Schacter and Moscovitch's (1984) review of the literature they conclude that the evidence supports the interpretation that infants show evidence of memorial functions long before they have anything remotely resembling a declarative, explicit memory. Moreover, the implication is that it is this implicit memorial system that survives if later neuropathologies cause amnesia. However, there is only a modest literature that looks at implicit cognitive functions over extended numbers of years of life, although the studies that exist support the general ontological arguments entailed by the evolutionary model. In a recent study Parkin and Streete (1988) used a picture-priming task with subjects ranging from ages 3 to 20. No differences were found between groups whether they were tested after delays of an hour or two weeks. Carroll, Byrne, and Kirsner (1985) reported a similar pattern of results using a picture priming task and subjects ranging from five to nine years of age. Greenbaum and Graf (1989) used both an implicit task (word production) and an explicit task (recall) and found no differences between three, four, and five-year-olds on the implicit task but significant improvements over this age range on the explicit task. Although this last study looks at only a small age range, it has the advantage of showing a dissociation between the implicit and explicit functions.

Fourth, there is a small but growing literature on very young children and infants that suggests that they are able to pick up on the true covariations in their environment and exploit them behaviorally. Haith and his co-workers (Haith, Hazan, & Goodman, 1988; Haith & McCarty, 1990) have shown that infants as young as 3 months pick up on alternating patterns in a stimulus display and begin to anticipate upcoming events. In these studies infants are presented with a visual display in which a face alternates from side to side. Initially the infants follow the appearance of each face; however, after several observing alternations the infants begin to make anticipatory eye movements to the location where the face will appear next.

Rovee-Collier and her colleagues (see Rovee-Collier, 1990, for a review of several decades of work) have presented evidence that young children quickly pick up the relationship between their own motor movements and the environmental impact that they have. In these studies the infant's foot is attached to a mobile that hangs over the crib and hence moves whenever they move. The infants quickly establish operant control over the movements of the mobile. By carefully controlling the contingency between the infants' movements and the actions of the mobile Rovee-Collier and her co-workers were able to firmly establish that the link here is deliberative and not merely adventitious.

Finally, there are two unpublished studies from our laboratory showing that children as young as four are capable of true implicit learning akin to that found with adults. In the first, Roter (1985) used a modified form of the artificial-grammar learning experiment and ran several groups of children with average ages of 4, 8, and 14 years and a group of college-aged adults. The youngest children in her study were able to learn the underlying structure of the grammar and showed no appreciable differences in behavior from the other subjects. The group that showed the most idiosyncratic behavior was, oddly, the preadolescents, who tended to engage in more explicit type learning than the others. In the other, children between the ages of 3 and 5 were run on what we call the "catch Max" game. In this task a cartoon character named Max appears at one of several marked locations on a computer monitor. The child's task is to "catch Max" by pressing a key that corresponds to Max's location. On some days Max's locations were determined by a complex set of rules, on others they were random. Over a four-day period the RTs to the rule-governed locations systematically decreased while the RTs to the random locations showed no improvement. In both of these studies young children were able to exploit the structure of their environments and make functional and adaptive adjustments in their behavior.

Of all of the entailments of the evolutionary model of implicit learning, this is the one most in need of serious research. There is a genuinely paradoxical quality about the manner in which this issue of implicit learning and cognitive development has been dealt with in the technical literature. It's like that old line about "the fish being the last to discover water." Since virtually everything interesting that a child learns about his or her social, cultural, familial, physical, and linguistic environments is acquired without support from conscious strategies for acquisition, the question of the interaction between implicit induction routines and explicit encoding rarely arises. There are research programs that look to explore such processes as a child's metacognitive capacities (see Flavell & Wellman, 1977), and of course within the Piagetian approach, a child's capacity to solve particular classes of problems is often compared with verbal descriptions of the nature of those problems. But, for the most part, the problem of implicit learning in children has not been perceived as one in need of serious research.

The disparity between the neglect of this issue among developmentalists and the continued fascination with it in the more general cognitive sciences is difficult to understand, both from a purely theoretical point of view as well as from an applied and practical perspective. For example, there are good reasons for suspecting that many of the thorny problems that crop up in the theory of instruction and many of the practical problems that emerge in pedagogic practice are the result of a failure to distinguish the implicit from the explicit. Most of our formal education is handled as though the acquisition of complex knowledge were an explicit, conscious process, despite the fact that we know that much of the critical knowledge that a child must bring to an educational setting was acquired implicitly. Some of these issues are discussed further below under the subsection on IQ independence and again in the next chapter.

Variability

The main point here is relatively straightforward. If implicit systems are evolutionarily old, they should show greater stability than the more recently evolved explicit systems. Since this kind of stability is reflected by "tight" distributions in a population, we would expect to find fewer individual to individual differences in performance between people when implicit, unconscious cognitive processes are recruited than when explicit, conscious ones are invoked.

Like the preceding issue, there is precious little empirical work on this problem. So far as I know, the only study carried out specifically on this problem is one by Reber, Walkenfeld, and Hernstadt (1991). In this study, all subjects (college undergraduates) worked with both a standard implicit learning task and a highly explicit task. The two tasks had superficial similarities in that both used sequences of letters as the stimuli. The implicit task was the standard artificial-grammar task in which subjects memorized rule-governed letter strings that had been generated by an artificial grammar (e.g., see Figure 2.1) and were tested for their tacit knowledge of the rules using the well-formedness procedure. The latter was a problem-solving task in which they had to solve alphabetic series problems based on rules for letter ordering, for example, ABCBCDCDE___ (D or C?) and CDEADCA___ (E or D?). (D is correct for the first problem, which is based on a simple "step" up the alphabet; E is correct for the second, which is based on a reversal rule involving two-letter groups.) Compared with the artificial-grammar learning task, this series solution task compellingly recruits explicit, hypothesis-testing behavior from virtually all subjects.

The critical comparison in this study was that between the probability of a correct classification response on the well-formedness task and the likelihood of a correct solution on the series solutions task. Each task used a two-alternative, forced-choice procedure, and so each yielded performance data that could be presented as the proportion of correct responses, P_c. The mean P_cs were .609 and .611 on the implicit and explicit tasks respectively and were statistically indistinguishable from each other. However, the variance on the explicit task was over four times as great as it was on the implicit task ($F_{max} = 4.54$; $p < .01$). This is but a single study, but the results are in keeping with the predictions of the evolutionary perspective.[7]

Although this is the only direct look at this issue that I am aware of, there are several other studies that provide supportive evidence. For example, in a rather different experimental context, Aaronson and Scarborough (1977) found an intriguingly similar pattern of results. Their study involved comparisons between several models of sentence encoding and used a serial reading task in which words appeared one at a time on a computer screen. Under such conditions two, distinct, independent processes emerged; one involving the reading of the individual words as they appear on the screen and organizing them into phrase units, the other the integrative process of encoding the sentential information. In keeping with standard psycholinguistic analyses, Aaronson and Scarborough regarded the former as a conscious, overt process and the latter as an automatic, unconscious one. The relevant data here were the individual differences observed on the two tasks. When slow readers (bottom quartile of their sample on overall reading scores) were compared with the remainder of the sample, large and significant individual differences were found on the explicit process of reading and organizing individual words with slow readers taking nearly three times as long to carry out this part of the task. However, no differences appeared on the implicit task; slow readers carried out the integration task at the same rate as the other subjects.

[7] It should be noted that some very interesting and difficult psychometric problems emerge when attempting to deal with the issue of comparing the variability of specific samples. In some ways, comparing the variability of implicit and explicit processes is like comparing the variability of, say, height and weight; without a common metric the question is mathematically meaningless. The comparisons carried out in this particular study involved a bit of statistical legerdemain based on the following four considerations: (1) each subject was run through both procedures, with order counterbalanced; (2) the tasks have superficial similarity in that each is based on rules for symbol sequencing; (3) performance was evaluated using two-alternative procedures so that both response measures are based on binomial distributions; and (4) the comparisons were "theory-driven" and were only carried out because the evolutionary model argued that such comparisons are of interest. There are other ways in which the low variability predictions of the evolutionary stance can be tested involving rank order data and the use of subjects of different ages; these are currently being pursued.

Also of relevance here are several studies of Snodgrass and her colleagues on fragment-completion priming tasks (Hirshman, Snodgrass, Mindes, & Feenan, 1990; Snodgrass & Feenan, 1990). In a series of studies they reported strong evidence to support the argument that the fragment-completion priming task is governed by two processes, one sensitive to explicit recall performance and one sensitive to conceptual representations in implicit memory. In unpublished analyses of the data from these papers (Snodgrass, personal communication), large individual-to-individual differences in performance were observed on the explicit phase but relatively few such differences on the implicit. Snodgrass has noted that the results of these analyses were not included in the published papers because she knew of no theoretical basis for either expecting or interpreting them and her research team regarded them as merely an interesting anomaly.

Some additional support for the variability proposition can also be derived even more indirectly, as follows. It is well known that the amount of variability in task performance increases dramatically with increasing age. Heron and Chown (1967) reported the shifts that take place across the life span on a standard measure of nonverbal intelligence (Ravens' Progressive Matrices Test). They found the typical gradual drop in mean performance, but they also found a large increase in the variability of scores with increasing age. There are many reasons for this shifting pattern of performance across the life span, the most commonly suggested (Shimamura, 1990) is that there is a large number of separate elements that make up the cognitive capacity being evaluated by such a test of intelligence and, like the parts of an old car, they are "breaking down" at uneven rates. Put simply, during aging the likelihood of any particular cognitive function becoming impaired is increased. Hence, variability will go up dramatically even though mean performance levels decline only moderately.

Note, however, that the tasks that have been used in these studies are typically highly explicit tasks, like the Ravens', which is a test of intelligence. It should be clear by now that implicit tasks do not function like explicit tasks; they are more robust and they do not show the same age-related shifts in performance. It is, therefore, not unreasonable to infer that the shifts in performance on implicit tasks across the life span would show a different pattern from the explicit tasks. Specifically, they can be expected to show less variability and a slower diminution in overall performance. Collecting such a data base is a major undertaking but it needs to be done.

Admittedly, the data base here is a bit impoverished. The reason for the lack of research on the variability issue is simple: no other model of implicit cognitive systems predicts it so there has been no motivation to explore this issue. It is, however, an important issue particularly because of the manner in which it dovetails with the following.

ON WHICH DIFFERENCES MAKE A DIFFERENCE. Before going on to the next issue, there is one additional point that needs clarification if we are to prevent confusions about what exactly is entailed by the prediction of low individual-to-individual variation. The evolutionary model predicts that implicit processes, being subserved (as I am assuming) by structures of considerable evolutionary age, will show less variability within any given population than explicit processes, which are dependent (as I am assuming) on more recently developed cortical features. Hence, nothing in this characterization argues against the emergence of even relatively large individual differences in the *explicit* cognitive components that surface in the typical experiments carried out with adults, such as those discussed above.

In several studies that have been carried out on implicit learning, rather large individual differences have been reported. For example, Dulany et al. (1984) reported that subjects in their artificial-grammar learning study tended to induce what they called personal "correlated grammars" and that each subject's was detectably different from the others. Similar findings were reported by Mathews, Buss, Stanley, Blanchard-Fields, Cho, and Druhan (1989), also using an artificial-grammar learning procedure. In both of these papers, the conclusion was reached that there were significant individual differences in implicit learning.

However, as was pointed out in the previous chapter, these differences tended to be ones that resulted from variations in the form of the mnemonic devices used by subjects during the memorization phase of the experiments; they are differences that, in terms of the utilization of tacit knowledge, do not make a difference. In our laboratory we have also observed and reported on such individual-to-individual variation (see Reber & Allen, 1978). However, we have noted that such encoding diversity does not necessarily point to any overall differences in performance on tasks that require that this implicit knowledge be put to use. Moreover, in these studies we do not know whether there would have been even greater subject-to-subject variability in performance had a highly explicit task been included. With the exception of the Reber et al. (1991) experiment, none of these studies has involved any direct (or even indirect) comparison between explicit and implicit functions. Such comparison are, of course, needed in order to examine properly this question of intersubject variability.

The deep issue here is the degree to which the underlying memorial representation of the stimulus display is coordinate with that of the actual structure of that display and not whether eccentric encoding schemes were a part of the induction of that representation. The divergences in behavior observed during the acquisition phase, considerable as they may appear, are not predictive of the subjects' abilities to use the tacit knowledge base to make decisions, judge novel stimulus instances, or categorize exemplars. When overall performance on the testing phase is examined, the individual-to-individual differences observed during acquisition are not found.

Other examples of this comparison are found in reports by Abrams and Reber (1988) and Rathus, Reber, and Kushner (1990) and the Reber et al. (1991) paper. In the Abrams and Reber (1988) study, discussed earlier under the *robustness* issue, psychotic inpatients and normal controls were found not to differ in their performance on their ability to discriminate well-formed from ill-formed letter strings. Recall, however, that the groups *did* differ in the number of errors it took to reach criterion on the memory task used during learning. The inpatients were clearly deficient in a number of the overt and largely conscious skills routinely found in normal subjects in these studies. For example, the use of the kinds of mnemonics that we, Mathews, and Dulany have found assist normal subjects in memorization, classification, and observation tasks were often missing and, when present, predictably bizarre. The inpatients also displayed slower response times and took much longer just to push the proper keys on the computer. However, once they reached criterion on the memory task, their performance on the well-formedness task was statistically indistinguishable from that of the normal controls.

A similar, although less dramatic, pattern of results was found in Rathus et al. (1990). The subjects were college undergraduates, who were divided into two groups depending on whether they scored above or below the median on Sarason's Test Anxiety Scale (Sarason, 1978). High-anxious subjects, like the inpatients in Abrams and Reber (1988), took significantly longer to memorize the learning items than the low-anxious subjects. However, on the well-formedness task the performances of the two groups were statistically indistinguishable, suggesting that the interfering effects of anxiety were felt only on the explicit task. Finally, a similar pattern of results was observed in the Reber et al. (1991) study in which marked individual differences were found during the memorization phase of the grammar-learning study despite the lack of such differences in the utilization of the tacit knowledge to make decisions about the grammatical status of items.[8]

The point is that there are, of course, explicit aspects in the learning phase of these experiments. Subjects do build "correlated grammars," they do key on different aspects of the stimulus array, and they do introduce idiosyncratic mnemonic devices. But, so long as these "personalized" aspects do not introduce distortions into the representation of the deep structure of the display, they will tend to result in tacit knowledge bases that yield surprisingly little in the way of individual variation in the capacity to use implicitly acquired knowledge. There are differences that make a difference and there are differences that do not. The idiosyncratic aspects of the initial phases of the learning process are in the latter category.

Finally, in one of our own studies, we did report clear evidence of individual differences in implicit learning (Kassin & Reber, 1979). In this study subjects were run through a standard artificial-grammar learning experiment. All subjects were treated identically although for statistical analyses they were divided into two separate groups based on a median split on their scores on Rotter's locus of control scale. We found that the "internalizers" were indeed better on all phases of the experiment than the "externalizers." However, this finding is most likely viewed as providing insights into the role of attention and the allocation of attentional resources rather than individual differences. The point is, as we argued in the original paper, internalizers are known to favor a cognitive style in which

[8] The individual differences observed on the learning stimuli are, however, psychologically interesting and are discussed in more detail in the next section.

the external world is attended to closely and the varieties of the stimulus displays are carefully monitored. The point is that the internal control component that typifies such individuals derives from the need to have as much information as possible about the world about so as to be able to make more carefully judged decisions when they are needed. Hence, people who score high on the internal side of the internal/external dimension tend to pay much more attention to the stimulus displays than those who score on the external side. There is a growing body of evidence that attentional factors play an important role in implicit processes (Cohen, Ivry, & Keele, 1990; Dienes et al., 1991), which fits with this interpretation.

IQ independence

This issue develops around several classic problems. To wit: What is "really" being measured by the standard tests of intelligence? What is the relationship between these measures, typically given as IQ scores, and "intelligent" behavior? Are there reasons for suspecting that the ways in which we attempt to come to grips with these questions will break along the lines of implicit/explicit cognitive functions? And, of course, can these various elements be fit into the evolutionary framework?

There is a good bit of recent work on intelligence and the techniques used for its assessment that is relevant to these issues, specifically that of Robert Sternberg and his co-workers (Sternberg, 1985, 1986; Wagner & Sternberg, 1985, 1986). Sternberg has developed a triarchic theory in which he argues that three critical subtheories are needed to form a complete theory of intelligence. The subtheories emphasize the importance of taking into account factors such as contextual or cultural biases, the ability to carry out automatic functions so as to be able to deal efficiently with novel inputs, how basic cognitive processes such as encoding operations are organized, and how new procedural knowledge is acquired. Sternberg argues that people are often more successful in real-world settings than would be expected if only their IQ scores from standard tests, which tap primarily conscious and overt abilities, were taken into account.

Several recent studies lend support to this point of view. Ceci and Liker (1986a, 1986b) and Wagner and Sternberg (1985, 1986) have both reported cases where there is little relationship between scores on standard IQ tests and ability to function in complex, real-world settings. These real-world settings range from horse race handicapping to the effective use of the kinds of tacit knowledge people generally pick up "on the job," which are associated with established criteria for professional success.

The thread that runs through this work is that tacit and abstract components of cognitive functioning operate largely independent of the overt and conscious and, as a result, do so independent of standard measures of intelligence. The link with the conceptualization of implicit learning is fairly obvious, and given the evolutionary framework, it follows that, compared with overt and explicit cognitive functions, implicit learning ought to show relatively little relationship with intelligence as measured by standard instruments.

The only study I am familiar with that sought specifically to explore these issues is the Reber et al. (1991) described above. In addition to running subjects through both the implicit and explicit tasks, they were all given four subtests from the Wechsler Adult Intelligence Scale-Revised (WAIS-R).[9] The performance scores from the explicit and implicit components of the experiment showed the predicted pattern of correlations with IQ scores on the WAIS-R. The correlation between performance on the explicit, problem-solving task and IQ was statistically significant ($r = .69$; $p < .01$), that between performance on the well-formedness test of the implicit grammar-learning task and IQ was non-significant ($r = .25$; $p > .05$). These two correlations were significantly different from each other ($p < .01$). And, not surprisingly, the two tasks did not correlate significantly with each other ($r = .32$; $p > .10$).

Additionally, in keeping with the theme developed above concerning individual differences in implicit and explicit tasks, there was a significant negative correlation between IQ and the number of errors made before reaching criterion on the learning phase of the implicit task ($r = -.48$; $p < .05$). This last finding fits with the notion that the artificial-grammar learning experiment, as mentioned above, has some explicit components.

Although this study, which had a relatively small sample size ($N = 20$), only yielded significant correlations between the explicit elements and the IQ scores, there are good reasons for suspecting that some of these smaller correlations with the implicit functions would become significant if the sample size were increased and the regression values remained the same. The important point, once again, is not that implicit tasks show no concordance with either explicit tasks or IQ but that implicit tasks are different in fundamental ways from the explicit—and that these differences can be seen in an evolutionary context.

It is clear that this line of research has some tantalizing implications for our general conceptualization of what it means for a person to display "intelligent behavior," for our understanding of how these "intelligent behaviors" are distributed throughout the population, and for theories of instruction and pedagogic method. In the next chapter, where I will allow myself some fairly unbridled speculation, these matters are discussed in more detail.

Commonality

This issue is one that has been a staple of animal psychology, comparative psychology, and ethology for well over a century. The proposition is simply that the basic principles of various psychological functions (in our case here *learning* is the primary one under consideration) can be shown to be equivalent in some deep way independent of the point on the phylogenetic scale where they are observed. So long as we are willing, as we have been since Darwin, to accept generalizations from animal work as providing basic insights into human func-

[9] The four subtests used were Picture Arrangement, Vocabulary, Block Design, and Arithmetic. These have been shown to correlate highly with the full-length version of the WAIS-R with values from .92 (Brooker & Cyr, 1986) to .96 (Doppelt, 1956) reported.

tion, we have implicitly accepted the validity of this principle. The only thing I wish to add to this well-argued perspective is the suggestion that the process we have all been assuming lies at the core of our generalizations is the process of detection of covariations or contingencies.

In what follows, I will roughly adhere to the general line of theorizing put forward by Rescorla and his colleagues (see Rescorla & Wagner, 1972, for the general contingency model and Rescorla, 1988, for an overview). This position has at its core a simple basic principle that organisms key on the covariations between events and, hence, learn to take advantage of the cuing function that emerges when some events are contingently associated with other events. So long as some stimuli in the environment are arranged so that their occurrences cue the occurrences of other stimuli, they will acquire statistical predictive power.

The Rescorla-Wagner model has distinct explanatory advantages over earlier accounts of conditioning, which emphasized the *co-occurrence* between the conditioned and the unconditioned stimuli. There are inherent shortcomings in this pure co-occurrence model. To take everybody's favorite example, why did Pavlov's dogs not come to salivate to the sight of Pavlov? After all he was present every time food was delivered, so why did the dogs not treat him as a conditioned stimulus like they did the bell? The answer, of course, is that Pavlov was also present every time food was *not* delivered and so he had no predictive validity as a cue for food. Only the bell, by virtue of the established contingency, possessed that statistical advantage. That is, simple co-occurrence is not sufficient to establish learning.[10]

Virtually all of the microanalyses that have been carried out on the data from the many experiments on implicit learning point to this general conclusion. Subjects learn to group and "chunk" stimulus elements according to their covariational patterns (Reber & Lewis, 1977; Servan-Schreiber & Anderson, 1990).

[10] There is, naturally, controversy over this proposal. Some (Gibbon & Balsam, 1981; Jenkins, Barnes, & Barrera, 1981) have argued that temporal relationships are the critical components of learning and not mere stimulus contingencies. Others, particularly Miller and his co-workers, have put forward the case that simple co-occurrences between stimulus events are encoded as well as covariations (Miller, Kasprow, & Schachtman, 1986; Miller & Matzel, 1988). They maintain that virtually all co-occurrences between stimuli are detected passively, although they may not be displayed in behavior unless particular circumstances occur that tap this (implicit?) knowledge.

The research on this problem has been carried out within the scope of animal learning and conditioning but, in keeping with the general theme here, it has messages of a fairly sophisticated philosophical kind that connect with the cognitive issues we have raised. A thorough analysis of these problems requires a separate book; however, let me point out that Miller's position must be taken seriously for the following reasons: First, in order to detect the statistically meaningful covariations in the environment, an organism must be encoding all of the co-occurrences so as to determine which have functional predictive value and which do not. Second, given this process, some kind of filter or inhibitory mechanism is needed to keep the unwarranted co-occurrences that have been encoded from influencing behavior. Third, this entire analysis invites a renewal of classic Humean arguments concerning the nature of causality and its epistemic status. After all, causality is "merely" the apprehension of contingency. These points are pursued further in the next chapter.

The tendency to detect these statistical covariations has emerged in studies using a broad range of patterned stimuli. Experiments have been conducted using sequences of targets (Kushner et al., 1991; Lewicki, Czyzewska, & Hoffman, 1987; Reber & Millward, 1968; Stadler, 1989; Willingham, Nissen, & Bullemer, 1989), rule-governed letter strings in synthetic languages (Reber & Lewis, 1977), and structured control systems in simulated production and manufacturing plants (Berry & Broadbent, 1984; Broadbent, FitzGerald, & Broadbent, 1986; Stanley et al., 1989). In short, things that occur together are perceived as forming the foundations for the induction of the tacit knowledge of the patterns of stimulation inherent in the environment. Just as Hasher and Zacks (1984) had argued that there was a general automatic encoding of the frequency of events in the stimulus world, so these studies suggest that there is also an automatic encoding of the covariational patterns of these events.

The interesting evolutionary question here becomes how has this acquisition function become modified phylogenetically while still maintaining an identifiable core process? After all, the kinds of procedures that yield the induction of a rich abstract structure in the experiments with artificial languages seem to bear little resemblance to the procedures needed to get a conditioned gill-withdrawal response in *Aplysia*[11] or even those used by Pavlov and Rescorla with dogs and rats. The point is that the commonality principle holds at the deep level of the detection of covariation. What a phylogenetic analysis reveals is that, with increasing neurological sophistication, organisms become capable of detecting more and more tenuous covariations. Organisms as primitive as *Aplysia* require unambiguous and nonvarying pairing of stimuli for learning; humans are capable of the detection of much more subtle statistical covariations involving many stimulus elements and environmental properties. Or, if one prefers other terminology, humans are capable of learning "rules."

If this argument holds, there will be no need to introduce any distinctly different mechanisms for acquisition as one looks at the fundamental principles of learning up and down the phylogenetic scale. The obviously sophisticated capacities of humans relative to more primitive species will be explicable in terms of the capacity to encode covariations with lower and more complex contingencies and weaker statistical predictive properties. As stated earlier, implicit learning is a primitive process with a long evolutionary history, and there is no reason for assuming that it ought to have been lost during speciation.

Finally here, there are some interesting reasons for suspecting that this basic and primitive process presumed to be at the core of implicit learning may be captured by some recent formalizations of connectionist modeling. Although this certainly is not the place (and I certainly lack the expertise) to put forward formal connectionist models of implicit learning, I would like to take a paragraph or two here to engage in a bit of speculation concerning their applicability to the

[11] A marine mollusk of rather modest intellectual accomplishments that, nevertheless, shows clear evidence of both simple classical and differential aversive conditioning (Carew, Hawkins, & Kandel, 1983).

general issues under discussion. One of the features of connectionist theory is that it is predicated on a fairly trivial process whereby inputs are encoded and represented in a massively parallel distributed fashion by nodes and the patterns of interconnectedness between the nodes (McClelland & Rumelhart, 1986; Rumelhart & McClelland, 1986). The richness and subtlety of these models is that they manage to establish their representations without resorting to specific rule systems; the rule-governed nature of the data base emerges from the patterns of covariation they encode.

In recent years a number of such connectionist models have been applied to implicit-learning tasks with considerable success. Jennings and Keele (1991) and Cleeremans et al. (in press) have successfully modeled the simple sequence learning experiments of Cohen et al. (1990). Cleeremans and McClelland (1991) developed a connectionist model based on a simple recurrent network (SRN) and applied it to a rather complex sequence learning technique in which the sequences of events were determined by the probabilistic properties of an artificial grammar. And Kushner et al. (1991) showed that the same SRN is sufficiently general so that it can be applied to a complex prediction task in which subjects must predict the location of a target event when that event is predicated on a complex, biconditional rule.[12]

It is certainly intriguing that these connectionist models, which have so much appeal that they have attracted the attention of animal behaviorists, neuropsychologists, philosophers, workers in artificial intelligence as well as conventional cognitivists, may be able to provide a formal foundation for examining the underlying process that we have long felt lies at the core of implicit learning.

Summary

I have tried, in this chapter, to present a somewhat different perspective on the cognitive unconscious generally and implicit learning specifically. The essence of my argument is that the invocation of some simple and relatively noncontroversial principles of evolutionary biology and an attendant functionalist stance leads to some surprising reformulations of a good bit of the data and theoretical work in contemporary cognitive psychology. I have also tried, where possible, to maintain contact with the traditional issues of epistemology as they pertain to the psychological questions raised by this orientation. Admittedly, portions of the preceding have a speculative quality, particularly in cases where the kind of

[12] It has recently been discovered that the SRN and other recurrent networks such as those proposed by Jordan (1986) have limitations when dealing with tasks of this type (Cleeremans, in press-a, in press-b). Interestingly, a simple decay-based buffer network appears to be able to account for the human data rather well. The reason this is so interesting is that buffer models of this type are rather simple networks, especially when compared with the SRN, yet they appear to simulate the data rather closely. Moreover, they provide an intriguing characterization of the representation process as yielding memorial representations that are intermediate between abstract and instantiated. (See Chapter 4 where this question of mental representation is pursued in some detail.) For details on the buffer models, see Cleeremans (in press-a, in press-b).

extensive empirical foundations we would like to have in order to support the arguments is lacking; the age-independence and IQ-independence principles are the most glaring. As I said at the outset, this is the "sensible speculation" chapter. Nevertheless, I think that the arguments put forward here have strong predictive components to them and the entailments of these proposals can be put to test. What is intriguing is that many of these entailments are not easily derived from any other viable theory of implicit learning or unconscious cognition currently being entertained. As I understand it, that is what "good science" is all about.

In the next chapter I will extend some of these issues, perhaps beyond the "sensible" point. I will also introduce a number of related issues and explore how the general perspective taken on implicit learning dovetails with some of the classic problems of epistemology and cognition.

4. Implicit issues: some extensions and some speculations

This chapter is the most general and broadest of the four. In it I try to deal with a variety of specific topics that have emerged in the study of the cognitive unconscious. As will become clear, some of the issues of concern here are direct entailments of the work on implicit learning; others are topics that are of general importance in cognitive science independent of any particular focus. Some issues will be primarily methodological, some theoretical, others philosophical. Not surprisingly, this chapter will be somewhat less well integrated than the others; my aim is to try to touch on a variety of issues of interest and, as will become plain, these issues don't always hang together easily.

Some of the lines of argument raised in this chapter go a bit beyond the material presented in the preceding chapters. Since the previous chapter was dubbed the "sensible speculation chapter," I thought about subtitling this chapter the "extended speculation chapter" or the "let it all hang out chapter" or something along these lines. Indeed, in some ways, there are elements of the following that have some of these characteristics—one of the hallmarks of this kind of writing being that there are fewer references in this chapter than in the previous three. However, I suspect that the various points of view presented here will not, in the end, turn out to be all that extreme. It strikes me that much of what follows may be little more than a resuscitation of common sense—a position that, alas, often appears quite radical. As I noted earlier, the philosophical point of view I defend here is nothing more radical than a benign form of old-fashioned representational realism grounded on a functionalist-based physicalism. In these philosophically heady days, this is about as close to common sense as one can get.

Implicit learning or implicit memory

This initial unit is an uncomplicated one with a simple focus: It is a plea for future researchers to be cognizant of the interrelatedness of learning and memory, particularly implicit learning and implicit memory. One of the very first points emphasized in Chapter 1 was that this was to be a book about *learning*. While much of the research that forms the evidentiary basis for this work was being carried out on the problem of knowledge acquisition, a veritable bandwagon was rolling along exploring the related problem of implicit *memory* (see Lewandowsky, Dunn, & Kirsner, 1989; and Schacter, 1987, for overviews). As noted, the past several decades of work in the cognitive sciences were ones in which models of memory took precedence over models of learning. It is, therefore, not surprising that when an interest in unconscious processes developed, attention was directed primarily toward the implicit aspects of memory rather than of learning.

For over a decade the two research programs have, unfortunately, traveled parallel courses with precious little interaction. Indeed, if one were to construct a Venn diagram of the literature citations in these two related domains, the intersection would be very nearly the empty set, a point also made recently by Berry and Dienes (1991) who have attempted to bridge the two areas of study by identifying common elements. I am sure that there is some lesson in the sociology of science to be learned here, and for whatever it is worth, I confess to having been as guilty as anyone else in contributing to this Balkanization of the field.

Fortunately, some workers are beginning to suspect that this mutual neglect is not to our benefit. Roediger (1990) recently called for greater communication between those studying implicit memory and those investigating implicit learning, Nissen and her colleagues have begun to explore both acquisition and retention of implicit and explicit knowledge (see, e.g., Nissen, Willingham, & Hartman, 1989), Squire and his colleagues (Knowlton, Ramus, & Squire, 1992) have added the study of implicit learning to their broader research program on memory, and Sherry and Schacter (1987) have taken to referring to larger physiological and behavioral systems that include acquisition, retention, and retrieval mechanisms, thereby explicitly focusing attention on the interrelatedness of these processes. These are all progressive steps, particularly Sherry and Schacter's, in which some intriguing evolutionary arguments are presented in an attempt to show how distinct memorial systems may have evolved—although I can't help noting that they still called their multiple-function schemas "memory" systems and that the bulk of their analysis is focused on memorial representations.

The obvious point, of course, is that learning and memory are so intimately interconnected that in the long run, it is going to be futile to attempt to develop systematic characterizations of one without the other. There can be no learning without memorial capacity; if there is no memory of past events, each occurrence of an event is, functionally, the first. Equivalently, there can be no memory of information in the absence of acquisition; if nothing has been learned there is nothing to store. Yet, for reasons probably having more to do with accidents of history and a kind of unthinking affiliation with particular methodologies than anything else, we find ourselves approaching them as though they were independent functions.

As the most obvious case in point, in the research on implicit learning outlined in the preceding chapters an attempt was made to isolate the acquisition process from the retention and retrieval processes as much as possible. In effect, perusal of the work of myself and my colleagues, as well as that of Broadbent, Brooks, Cleeremans, Lewicki, Mathews, and the many others who have examined implicit *learning,* reveals that terms like "memory," "retention," and "storage" occur infrequently. These studies share common elements that, as will become plain, are not found in the studies designed to explore implicit memory. The most prominent of these was that in these experiments stimulus materials were used that were not part of the subjects' preexisting representations upon entering the laboratory. Rather than use standard word lists or other largely linguistic material, the studies of implicit learning used stimulus materials derived from such arcane sources as artificial grammars, stimulus sequences based on complex and often probabilistic rules for position and order, and simulated control settings involving imaginary manufacturing plants and mythical computer people with odd swings of emotion. Moreover, in these experiments subjects were typically given instructions that discouraged the use of explicit heuristics and overt problem-solving strategies. Since these studies were designed to examine the process of unconscious induction of abstract knowledge about a complex environment largely independent of overt and explicit cognitive processes, the question of how and when to explore implicit memory never really arose.

To be sure, the subjects who went through the acquisition phase in the typical experiments in implicit learning emerged with a rich memorial store—what we have often called, following Polanyi, *tacit knowledge*. Consequently, the studies reviewed earlier on the manner in which subjects used a tacit knowledge base to make decisions and solve problems could, if one were so disposed, properly be viewed as studies of implicit *memory.* Those of us who were studying implicit *learning,* however, made the acquisition phase central and used the memorial and retrieval processes as diagnostic. That is, the representation and retrieval data were taken as evidence of the truly implicit nature of the induction process and the unconscious nature of the epistemic contents of memory.

The recent examination of implicit memory has, of course, been burdened by the flip side of this methodological coin—virtually all the work has been carried out using preexisting knowledge. Because the primary interest was in the *retention* component of the system, little attention was paid to how knowledge was acquired and very different classes of stimulus materials were used. With few exceptions (e.g., Johnson, Kim, & Risse, 1985), the study of implicit memory has examined the capacity of subjects to provide evidence of some tacit memorial residue, not of newly acquired material, but of particular episodes of previously known material to which they had been exposed under varying, controlled conditions. In the typical experiment on implicit memory, subjects are presented with material that is unambiguously part of their previously established representations. The studies have typically used common words, word fragments, familiar geometric forms, and the like as stimulus materials. Hence, when *learning* takes place in these studies, it consists essentially of asking the subject to memorize lists of familiar items, form new associations between common words, respond to previously encountered visual displays, complete word-stem problems, and so on. The kinds of implicit learning effects reviewed in the previous chapters are unlikely to be observed under these conditions. In the implicit learning experiments, the acquisition processes were central; in these implicit memory experiments, there was little or nothing to be "learned." Implicit *memory* was the phenomenon to be studied and terms such as "learning," "induction," and "acquisition" are as rare here as their memorial counterparts are in the learning literature. Saddled with such methodologies, the examination of learning, in the sense of the acquisition of genuinely new knowledge based on complex and structured displays, rarely arose. In retrospect it is quite amazing, at least to me, how wedded we have become to particular methodologies.

Where these two research programs do overlap, at least in principle, it is in the process of *retrieval*. Unfortunately, the connection has been difficult to spot, partly because of the unhappy connotations of that term. In the recent *Dictionary of Psychology* (Reber, 1985), retrieval is defined as the process of "recall" of information from memory. Alas, *recall* carries with it a sense of a conscious and overt process. This need not be. We simply need to recognize that the process of retrieval can, and perhaps should, take on a more neutral sense, at least with regard to the degree to which it is controlled by consciousness. When a subject retrieves information from a memorial system he or she is "reaching in" as it were into that memorial system and tapping an epistemic source that will have some role to play in directing behavior. We have typically viewed such a "reaching in" as a conscious process, but as the work on both implicit learning and implicit memory shows, it is often unconscious. The difficulty lies in that this "reaching" metaphor connotes an overt and reflective act. One of the reasons for taking the evolutionary stance argued for in the preceding chapter is that it leads us away from this sense that the conscious is primary and allows us to recognize that the processes involved in retrieval can occur independent of conscious control and of awareness of the process itself.

Indeed, it should be clear that there are numerous circumstances under which a subject is requested to retrieve previously stored information, some of them are overt and conscious like recall and, to a somewhat lesser extent, recognition. Others are covert and based on representations held unconsciously, like the well-formedness decisions in the artificial-grammar learning studies or the lexical decisions in the repetition-priming studies of implicit memory.

In any event, one of the advantages of the approach I have taken toward implicit processes is that it becomes possible to see how both implicit learning and implicit memory share particular patterns and characteristics. By couching both within an evolutionary context and by focusing on the patterns of dissociation that are common to both, we should begin to appreciate the extent to which they share basic features.

On rules

Rules play a large role in the cognitive sciences. We argue that a subject's behavior displays sufficiently coherent patterns that suggest he or she is using rules; we speak of some forms of concept learning as being based on rule discovery; we characterize stimulus displays as being rule governed in the sense that formal procedures were used to generate them; and we introduce rules as explanatory devices in our theories of complex behavior.

It is becoming pretty clear that lurking behind these varied uses of the notion of "rule" are some messy epistemological questions, questions like What is a rule? Where is a rule? What does it mean to say that a subject knows or holds a rule? What kinds of evidence do we need to conclude the existence of rules or rule-governed behavior? In the sense I am going to use it here, basically a rule is a theoretical device introduced by a cognitive scientist to provide some kind of explanatory handle on the patterns of regularities that can be found in various components of an experiment. The various uses of these rules can be codified into one of three *types* or, better perhaps, *locations*. Within the traditional experiments in cognitive psychology, like those we have been concerned with up to now, there are rules that are "in the subject," rules that are "in the stimulus display," and rules that are "in the cognitive scientist who functions as the experimentalist and theoretician." Within the study of implicit learning each of these uses has led to problems primarily because any rule that is going to have any serious explanatory power is going to have to be manifested in a coordinated manner in more than one of these locations, and this has often not been the case.

Let's take what I suspect is the signature case here in the study of implicit learning and look at it more closely. This case occurs when the cognitive scientist constructs a stimulus environment according to particular formal characterizations. During the course of the experiment, when the subject comes to behave effectively in the presence of this stimulus environment, he or she is presumed to know the rules that generated the display (in some sense of the word "know"). In a standard artificial-grammar learning experiment, for example, the assumption is made that since the stimulus display consists of letter strings generated by a finite-state, Markovian process, the subject, whose behavior is consistent with the structure of these letter strings, must have induced that grammar, at least partially.

As will become clear, this characterization of this experimental setting is seriously flawed by an inherent uncertainty principle, and this is probably a good time to "fess up" as the one who first committed this blunder (Reber, 1965, 1967a, 1969). My error was failing to take into account the question of coordination between the three places where rules can reside and is responsible, although only in part, for some of the disputes over the nature of implicit learning I and my colleagues have had with Don Dulany and his co-workers (Dulany, 1991; Dulany, Carlson, & Dewey, 1984, 1985; Reber, Allen, & Regan, 1985) and Pierre Perruchet and his colleagues (Perruchet & Pacteau, 1990, 1991; Reber, 1990). Parallel confusions, as will become apparent, have also been an element in the dispute between Perruchet, Gallego, and Savy (1990) and Lewicki and his co-workers (Lewicki, Czyzewska, & Hoffman, 1987; Lewicki, Hill, & Bizot, 1988) as to the true nature of implicit learning experiments carried out using complex sequence-learning procedures.

The point here actually turns out to be fairly simple. Just because a cognitive scientist uses some formal device to generate a stimulus display (as, for example, the Markovian systems we routinely use in our artificial-grammar learning studies) does not mean that any select subgroup of exemplars that comprises that display (or, for that matter, any other arbitrary group of exemplars up to and including the full set generatable by the grammar) is necessarily uniquely characterized by that formal device. For example, in an early paper on the learning of an artificial grammar (Reber, 1967a), I argued that subjects' knowledge base should be viewed as a partial but representative subset of the rules of the grammar. What I should have argued was that the subjects' knowledge base could be viewed as a partial but representative subset of the patterns among the various elements of the stimulus display that are reflected in the environment. The error was in assuming formal equivalence between, on one hand, the rules that were used to generate the display and, on the other, the rules that characterize the order in the stimulus display and (even worse) in presuming that these two characterizations can be taken as equivalent with the rules that subjects can be presumed to have induced.

This nonequivalence became especially clear in a later learning study (Reber, Kassin, Lewis, & Cantor, 1980) in which subjects were presented with the full formalization of the grammar either before, partway through, or after the completion of an extended learning phase. This presentation consisted of showing each subject the schematic diagram of the Markov device, like those in Figures 2.1 and 2.2, and explaining how it was used to generate strings of letters. According to the naive view here (i.e., the view that assumes equivalence between the formalization, the instantiated display, and the subjects' memorial representations), the best performing group should have been the one given the full formalization at the very end since this procedure would merely be concretizing a mental representation that had already been established during the learning phase. However, this group turned out to be the poorest of the three and for the simplest of reasons; subjects were using complex coding systems based on the patterns of covariations between letters, which had become formed into two- and three-letter groups or chunks to represent the stimuli. The sudden "insight" that the stimuli they had been working with were "in reality" sequences of letters based on an underlying finite-state system was not only of little or no value, it actually tended to depress performance below that of subjects who had had less experience with the stimulus display prior to the formal explication of structure. The point is, of course, that the further into the learning phase a subject was when the explication of the formal system was introduced, the more alien it was perceived to be in terms of the mental representations that the subject had been inducing. The key is the recognition that, while the cognitive scientist may make up a display using rules, the subject sees the display only, not the rules.

A similar problem was recently discovered by Perruchet et al., (1990) to exist in the work of Pawel Lewicki on implicit learning of rule-governed stimulus sequences. In a series of experiments Lewicki and his colleagues (Lewicki et al., 1987; Lewicki et al., 1988) had presented subjects with a complex sequence-learning task in which they had to react to each successive appearance of a target stimulus as quickly as possible. In the standard design (see Lewicki et al., 1988) the target could appear in any one of four equal-sized quadrants on a computer monitor. The stimulus sequence was divided up into logical blocks of five trials with a complex rule system determining the location of the last three targets. The rule system was based on having the first two locations determined pseudorandomly (the only constraint was that they could not both occur in the same quadrant) but having the last three locations determined by the locations of previously seen targets. Locations were determined by second-order recurrent rules that depended on the two previous locations within each block of five trials; specifically, if the target moved horizontally from Trial 1 to 2, then the move from 2 to 3 was vertical; if the move from Trial 2 to 3 was horizontal, then the move from 3 to 4 was diagonal; and so forth.

Over several thousand trials, subjects showed systematic reductions in RTs to the appearances of target stimuli on the three rule-governed trials relative to the RTs for the first two, pseudorandomly determined trials. Moreover, when the rules were abruptly changed, subjects showed a sudden rise in the RTs on the rule-governed trials but no changes in RTs on the pseudorandom trials. These results strongly suggested that the subjects had induced the patterns in the display. Interestingly, the subjects in these studies, who ranged from naive undergraduates to professorial colleagues of the authors who were familiar with the thrust of their research program, were unable to describe the rule for stimulus order they seem to have learned, even when offered a $100 inducement.

In designing the stimulus displays, Lewicki and his colleagues, unfortunately, had made a methodological error parallel to the one that we had made in our earlier work. There was, as Perruchet et al. (1990) pointed out, more than one way to capture the patterns in the stimulus display. While the rule that governed the location of the target stimuli on the critical trials was in conformity with the rule Lewicki et al. had developed, it was not the only rule that captured the structure of the display and, as Perruchet et al. (1990) were able to show, it was also not the one that most of the subjects were using.

The problem was that the manner in which Lewicki et al. set up the movement rules for successive target locations led to differential frequencies, not of simple locations of targets, but of patterns of movements of targets. The movement patterns turned out to be unevenly distributed over Trials 1 and 2 compared with Trials 3 to 5, with the infrequent movements occurring mainly in Trials 1 and 2. A simple frequency effect, therefore, would predict that slower RTs should occur on Trials 1 and 2 than on Trials 3 to 5. Detailed analysis showed that a variety of specific movement patterns had very different frequencies in the two types of trials. By analyzing the microstructure of their RT data, Perruchet et al. were able to provide convincing evidence that subjects were unlikely to be using the rules that Lewicki et al. had used to construct the display; rather they appeared to be responding primarily on the basis of the relative frequencies of movement patterns, with the more frequent patterns having the faster RTs. Hence, it comes as no surprise that Lewicki and his colleagues were able to keep their $100 rewards. In fact, it's not at all clear that they would have awarded it to a subject who replied with a rule based on relative frequencies of movement patterns.

However, it is important to note that this methodological artifact that Perruchet et al. identified does not have anything near the explanatory force they seem to think it has. While it rather poignantly exemplifies the problems that can emerge when one is unclear about the nature and location of rules, the discovery of this frequency artifact does not in any way undermine the original conclusions that Lewicki and his colleagues drew from their research; subjects are still engaging in implicit learning in these experiments. Perruchet et al.'s critique of the Lewicki procedure, accurate as it appears to be, tells us nothing startlingly new about the behavior of the subjects under these experimental conditions. There still is every indication that subjects in all of these experiments are learning a complex structure that characterizes a stimulus display relatively independent of awareness of both the process of learning and the knowledge acquired. The only change is in the characterization of the "rules" that characterize the stimulus display and, by extension, the subjects' mental representations. Lewicki et al. thought they could be captured by rules based on interdependencies between locations; Perruchet et al.'s analysis suggests that they were, in fact, based on relative frequency of movement patterns.

I point this out because Perruchet et al. (1990) do not seem to appreciate this side of the issue; they argue as though they had uncovered some fundamental flaw in the work of Lewicki and his colleagues. But this is simply not the case; they merely found a slightly embarrassing methodological glitch that Lewicki and his colleagues probably wish they had found themselves. The epistemological issue of interest to me here, however, is that it points out again the importance of the question of just "where the rules are." The general representational realist position is unaffected.

Lewicki and his co-workers, it seems clear, committed the same type of error we did in our earlier work, probably because he followed the same seemingly logical sequence we did. First, make up some rules to generate stimuli; second, assume that those stimuli actually are uniquely defined by those rules; and finally, assume that when the subjects "act right" with respect to the rule-governed display, conclude that they know *these* rules. Now, of course, this characterization may be true, but then again it is likely not to be. The only real rules in this three-level arrangement, of course, are those in the cognitive psychologist's head; the others are descriptions or, more accurately, *inventions* of the cognitive psychologist that capture, to some subjective or statistical criteria, the patterns reflected in the stimulus environment and the consistencies observable in the subjects' performance.

When misunderstandings like these emerge, it is typically because only one of the three "places" that rules can be has been specified. In order to avoid artifact in methodology and embarrassment in interpretation, the rules operative for at least two of the three levels must be properly represented in compatible form. For example, in both the sequence-learning experiments and the artificial-grammar learning studies, only one of the possible locations had an unambiguous rule, the cognitive psychologists' heads. With two degrees of freedom in a three-factor arrangement like this one, uncertainty is bound to leak in.

The point of all this is that there are no "rules" in the "real world." A kind of pragmatic metaphysics is needed here. Let's try out the following: The stance that makes sense is the one that assumes that the experimental cognitive scientist exists on a separate explanatory plane from the two objects of investigation—namely, the stimulus environment and the subjects' behaviors with their inferable cognitive processes. The only place that we can be sure a rule exists is in the cognitive scientists (i.e., in us) simply because we have access to our own rules and, as specified, the cognitive scientist here needs to be represented as existing on a different explanatory level.

When the cognitive scientist constructs[1] a stimulus environment, he or she may do so on the basis of some set of principles that have the effect of creating an environment that reflects particular patterns of co-occurrence and covariation among its elements. But there are no *rules* here, just patterns of co-occurrence and covariation. The cognitive scientist may *think* that there are rules that characterize these covariations, and in fact, she or he is certainly entertaining a particular clutch of these—namely, the ones begun with.

However, when subjects are exposed to these patterns of covariation under any of a number of possible scenarios and cover stories, all manner of things may happen. One of them, and the one that most interests me here, is that the subjects begin to modify their behavior in the context of the stimuli in ways that suggest a growing sensitivity to the patterns or co-occurrence and covariation. They may do this in any number of ways, for example, by displaying systematic decreases in their RTs, by becoming more efficient in memorizing structured symbol strings, or by coming to make appropriate categorization and classification responses concerning the structural status of novel stimuli.

Now, at this point, do we (the cognitive scientists) want to say that the subjects who perform so effectively in these complex settings "have rules"? There are two answers here and they are the usual two: Yes and No. Yes, there will be occasions when we will want to endow our subjects with mental representations based on rules on the grounds that their behaviors are clearly reflective of those of an organism who "has rules" and, in so far as this organism is detectably a conspecific, we feel it is legitimate to grant them the same capacity to "have a rule" that we have already accorded ourselves. But then again, maybe the proper answer is No. This conclusion may be the appropriate one under conditions where there are no other, independent reasons for going this interpretive route. It is, of course, possible that our subjects are merely responding, not to the patterns of covariation that we have taken as defining the stimulus display, but rather are setting up fragmentary, episodic, instantiated memorial bases and making decisions and judgments that have no rule-like cognitive basis to them.[2]

[1] While the concern here has been with cases in which the nature of the stimulus display is under the control of the scientist, a parallel argument holds in other cases as well. For example, in the case of explorations that involve the use of more natural environments or cases where naturalistic observation techniques are employed, the rule-governed nature of the display is induced from examination of the environment rather than imposed directly. The point of the uncertainty argument, however, still holds; any *formalization* that is put forward to characterize the patterns and consistencies displayed in the environment is the creation of the cognitive scientist and is not necessarily a unique characterization of the environment. This is, of course, one of the primary reasons why doing science is so difficult.

[2] This latter point is precisely the one that has been vigorously defended by Lee Brooks and his co-workers in a variety of papers (e.g., Brooks, 1978; Brooks & Vokey, 1991; Vokey & Brooks, 1992). There are some interesting elements behind this line of argument; they are pursued in more detail in the next section on knowledge representation.

How do we unpack these possible interpretations? Well, one way would be to ask the subjects to tell us which rules they are or are not using. When the knowledge is held explicitly and the subjects' descriptions of their mental processes and contents are consistent with both their behavior and with the structure of the environment, then we would be likely to concede that they do, indeed, "have rules." Recall that the factor that allows us to put the cognitive scientist on a separate level here is the simple fact that the cognitive scientist is consciously aware of the rule system that he or she has built into the stimulus display and can readily communicate it to others.

However, when the knowledge is not available to consciousness, the issues become tricky. We tend to be reluctant to regard our subjects as "having rules" when they cannot explicate such; after all, we only granted the cognitive scientist the right to have a rule because he or she could articulate it. We feel, I suspect, more comfortable with more conservative glosses like "the subjects' performance indicates that they have internalized a rule" or some similarly euphemistically laden line.

As a short aside here, it is worth recognizing that the subjects in these experiments are a lot like scientists making their initial approach to a new field of study. They have been given exposure to limited, representative samples of an arbitrary domain of reality. Initially they "process" information in a fashion that involves two distinguishable operations. First, they work with the stimuli, qua stimuli. That is, they initially confront the stimulus display via a set of standard practices and procedures concerning how they are to interact with them. In some cases they memorize them, in others they observe them, in still others they read them from a list, or form paired associates with them. These kinds of processes function to familiarize the subjects with the range of the stimuli and with the patterns of covariation that they display. Note that many elements of this aspect of the acquisition process are quite conscious; subjects typically use a variety of mnemonic devices during the learning phase of these studies, and they can frequently tell you what they were (see Allen & Reber, 1980; Reber & Allen, 1978). Moreover, the kinds of conscious processes and the types of mnemonics that subjects invoke have been shown to depend to a considerable extent upon the kinds of instructions they are given and the demands of the specific task (Reber, 1976; Reber & Allen, 1978; Reber et al., 1980). Second, subjects induce a memorial representation that reflects these patterns of covariation. This latter operation *is* the process of implicit learning. It is the process that takes place largely independent of those aspects of the conscious processes that are involved in memorizing or observing the stimuli. It is also the process that provides the subjects with a particular set of "rules" that can be used to characterize their mental representations.

It is this implicit induction process that the philosopher Michael Polanyi referred to when he described the essence of the work of the creative scientist as building up a "personal knowledge" that resisted verbalization but nevertheless was the driving force behind the ultimate attempts to come to understand the "knowable reality" that was out there (see Polanyi, 1962; and Reber, in press, for a discussion of the relationship between Polanyi's epistemology and the theory of implicit learning). Polanyi often referred to the state of "knowing more than we can say" about epistemic domains. In fact, he regarded this personal or tacit knowledge base as doubly refractory to consciousness, the knower being incapable of explicating what was known to others but similarly incapable of explicating it to himself or herself. Polanyi's entire sociology of science is predicated on the proposition that, in large measure, tacit knowledge is simply not open to conscious explication. That, he argued, is one of the reasons why doing science is so difficult. It is also, I suspect, why being a subject in a cognitive psychological study can be so difficult and why we must struggle to understand each.

Still, there is, I believe, a proper (and, of course, pragmatic) stance to take on this issue of rules, what they are and where they are. It involves the acceptance of the position of representational realism. That is, it assumes that a subject in an implicit learning experiment is inducing a mental representation that is a partial isomorphism of the structure inherent in the stimulus display; however, the subject's mental representation needs to be examined on its own merit as though it were but another mental state about which one may theorize. It is likely that insights into the nature of mental representation will be garnered from careful examination of the structured nature of the stimulus display, but it is also likely that the operative description of the array may not correspond with the initial formal characterizations that were used in its construction. In short, we know with surety what the rules were that we used to construct the displays. But we cannot know with anything like the same confidence either the rules that characterize the display nor those that are induced by our subjects. Careful examinations of their behavior will provide some understanding of the rule systems they have induced, and the basic principle of representational realism suggests that we regard this characterization as the one best suited for the stimulus displays themselves.

Finally, there is one more issue that needs discussion here, the question of how mental representations are captured within the various formal models of implicit learning that have been proposed recently. Some of these models are based on "rules," for example, the production systems of Holland, Holyoak, Nisbett, & Thagard (1986) and the classifier models of Druhan and Mathews (1989), and some are "rule-free," for example, the connectionist models like Cleeremans's simple recurrent network (Cleeremans & McClelland, 1991; Kushner, Cleeremans, & Reber, 1991). The rule-based systems have explicit production routines embedded in them, such as the classic "IF-THEN" procedure, whereby specific conditions call for particular actions (Holland et al., 1986), and the "forgetting algorithm," in which new rules are created from exemplars by retaining only subsets of features from the old ones (Druhan & Mathews, 1989; Mathews, Druhan, & Roussel, 1989).

There are two interesting features about these models. First, these two kinds of formalisms seem to be as different epistemologically as they are in their formal structure. Put simply, the connectionist systems seem to be "lower level" models, whereas the production systems feel like "higher level" models. That is, if we view them as classes of models that at least attempt to capture particular elements of human cognition, the associationist systems appear to be modeling more basic, even primitive processes than the production systems. Not surprisingly, the connectionist systems have been eagerly pursued by those with a neuroanatomical orientation with their focus on the notion of a "neural network." On the other hand, the models based on production systems tend to feel more like higher-level cognitive operations; that is, they seem to be capturing, not the underlying networks among elements, but *thought processes* themselves. Given these features, it seems pretty obvious that the two classes of models should be viewed more as complements than as competitors. In all likelihood, a more powerful model would emerge from an amalgamation of the two approaches with the connectionist systems modeling the lower-level associationistic functions and the production systems capturing the seemingly more complex, rule-governed activities. What is needed is the bridge, the set of operations that allows associationistic nets to be represented as rule-systems and algorithms. The parallel between the implicit and the explicit here is, of course, just as obvious. Hence, giving full rein to the speculation I promised in this chapter, I would like to suggest that connectionist systems will do a better job at simulating the implicit processes, whereas the production systems will be more successful in capturing the explicit.

Second, it seems to me that the issue concerning the manner in which these two classes of models can be said to have or not to have rules parallels closely the issue of determining whether or not our subjects in these experiments do or do not have rules. After all, the internal state of the computer system used to carry out the simulation in the case of these models is merely one more "place" where a rule can be.

Connectionist models may not have rules or symbols in that there are no rule-like properties in their operating systems and no symbolic content in the pattern of nodes in their networks, but they do have sets of connections and associative links that make them "act like" they have rules. Production systems, on the other hand, do have rules in that the structure of the model is predicated on rule-like operations or algorithms that move symbols about according to particular kinds of formal principles. Thus, from one perspective, the reason that a model like THYIOS (Mathews et al., 1989) has rules while one based on a simple recurrent network (Cleeremans, Servan-Schreiber, & McClelland, in press) does not is that the developers of THYIOS say that it has rules whereas the developers of the SRN say that it does not have rules. In a reductionist world, however, THYIOS's "rules" are reducible down to digitalized patterns of 1s and 0s just like the "non-rules" in the SRN machine, and both are just like our subjects' rules, which may be instantiated as patterns of neural firings (plus some really messy things like hormones). Could the algorithms of THYIOS be instantiated as an associationistic network? I would expect so. Similarly, I suspect the complex network of the SRN could be instantiated as a set of rules with symbolic content. If these translations were to be successful, where would the rules be?

As I see it, ultimately all of this comes down to matters of pragmatism. We have to decide, in advance, the level of analysis we wish to take. If we choose to work in the mode of the reductionist and aim toward the molecular, then there are no rules or symbols to be found anywhere—both the artificial, silicon-based connectionist and the natural, carbon-based neuroanatomical systems are devoid of rules in just the same way. To paraphrase a favorite phrase of the ontologists, "It's neurons all the way down."

On the other hand, if we choose to work in a somewhat less elementaristic frame then we may legitimately introduce rules and symbols. We, the cognitive scientists, work with rules and our subjects may indeed work with rules and so may our simulating machines. If some find it unsatisfying to use rules and symbols as the language of the science, that's perfectly all right. It doesn't mean they don't exist in some larger metaphysical sense, only that they have no explanatory role to play in those particular theoretical analyses. The essential issue becomes one of determining how and at what level one is best able to achieve insight into the system under analysis and articulate the model that has the greatest explanatory force. The one I choose is the pragmatist's stance, in which it is legitimate to talk about rules with the understanding that, of the three places where rules can be, we should only feel comfortable when we have coordination between at least two of them.

This issue leads quite naturally to several neighboring problems that are of interest. The natural segue here is to the question of the nature of the mental representation that subjects induce during implicit learning. In short, what do their "rule systems" look like?

Knowledge representation

One of the key theoretical issues in the study of implicit learning has been the nature of the cognitive content that a subject emerges from the learning phase with. In Chapter 2 a defense of "representational realism" was put forward. It was argued that although there were various ways an individual subject's mental representation could be characterized, the lion's share of the evidence was for the representation that reflected, albeit only partly, the underlying abstract structure of the display to which the subjects had been exposed. Here, I would like to take this position, which for the present purposes we will call the "abstractive" view, and compare it with several other theories of knowledge representation that have been argued for recently.

The abstractive view

This position is based on the argument that the complex knowledge acquired during an implicit-learning task is represented in a general, abstract form. The term *abstract* is a common enough term in psychology, but it is also a slippery one that can cause no end of confusion. The manner in which I use it here derives from its original Latin meaning of "drawn away from" (see Reber, 1985). That is, an abstract representation is assumed to be derived yet separate from the original instantiation. Abstract codes contain little, if any, information pertaining to the specific stimulus features from which they were derived; the emphasis is on structural relationships among stimuli.

Take, for example, a letter string from one of our artificial grammars, TPPPTXVS. An abstract representation of this string would be of the form "a single occurrence of letter-type 2, followed by three occurrences of letter-type 1, another single occurrence of type 2 and single occurrences of types 5, 4, and 3."[3] The essential argument of the abstractive view is that, with continued exposure to exemplars of this kind, subjects encode the display deeply and set up a representation that captures the patterns of covariation between the various stimulus types that are displayed. The key feature that differentiates this view from the others below is that the hypothesized mental content consists not of the instantiations of specific physical forms, but of abstract representations of those forms.

When subjects are asked to use this knowledge in some fashion, such as discriminating novel well-formed from ill-formed strings, they are assumed to compare the abstract coding of each of the novel test strings with the previously established deep representations and to judge well-formedness by imposing a criterion based on the degree to which these codings match. Now, there are some obvious difficulties with this view. For example, the issue of abstraction itself is, to put it mildly, poorly understood. Exactly how the various components of this kind of representation are coded with respect to each other is still an open question, although there are reasons for suspecting that the processes may be captured by connectionist models that build up representations from the patterns of covariations among stimulus elements (Cleeremans & McClelland, 1991; Kushner et al., 1991). Moreover, it is still far from obvious how subjects establish criteria that permit the comparison between the abstract encodings of the novel test stimuli and their memorial representations.

Nevertheless, the abstractive view has considerable explanatory power. For one, it handles the basic data in the field with aplomb, particularly the issue of how subjects deal with novel stimuli that are physically dissimilar from those experienced during learning. That is, the abstractive perspective is the only model of mental representation that can deal with the existence of transfer of knowledge across stimulus domains. Since this issue of transfer is so central to the question of knowledge representation, it will be dealt with in more detail below.

The distributive or exemplar-based view

This perspective on implicit learning and representation, which is referred to as either the distributive or the exemplar-based view, has been championed primarily by Brooks and his associates (Brooks, 1978; Brooks & Vokey, 1991; Vokey & Brooks, 1992), although the issues raised have also been argued to apply to

[3] For purposes of this example, the numerical ordering of letter types was handled alphabetically. This is, of course, arbitrary and any other convenient code would suffice.

more general issues of conceptual memory (Medin, 1989; Smith & Medin, 1981). The distributive view holds that a stimulus is coded and stored, not on the basis of patterns and regularities among features, but as a kind of "raw" instance. A string like TPPPTXVS is represented as "TPPPTXVS." The basic argument here is that with continued exposure to exemplars of this kind, subjects build up a distributed memorial store containing numerous instances of specific items. When called upon to use this knowledge in something like a well-formedness task, subjects are assumed to compare each new string with the instantiated memories and judge well-formedness by imposing a similarity criterion based on the degree to which each test string compares with the stored strings. Stimuli that have a physically close match with items in store are accepted as well-formed; those with no such match are rejected as ill-formed.

Here, too, there are problems with these approaches to the representation issue. For example, exactly how large numbers of specific instances are stored and kept distinct from each other is not well understood, and just how similarity between novel and stored stimuli is coded and judged is a problem that is inherent in approaches of this kind. Nevertheless, the exemplar-based, distributive view has considerable currency in the literature, and models of this kind have been applied widely in areas as diverse as social judgment (Smith & Zárate, 1992) and category learning (Kruschke, 1992). Exemplar-based models have an advantage over abstractive models in that the process of storing exemplars appears to be fairly mechanical and straightforward; no recoding of the stimuli is required and there is no need to carry out inductions based on patterns and structures reflected in the display. However, exemplar-based models have a disadvantage in that they are intimately tied to the physical form of the input stimulus and hence are inflexible.

The fragmentary view

This position is similar in some respects to the distributive, exemplar-based view except that less emphasis is put on the holistic properties of the exemplars. Theorists who defend the fragmentary perspective emphasize the role of more "micro" elements of the stimulus displays. Several of these fragmentary models of implicit learning have been proposed. For example, both Dulany et al. (1984, 1985) and Perruchet and Pacteau (1990, 1991) have suggested that in a typical artificial-grammar learning experiment each subject develops a "personal," correlated grammar within which letter groups are coded and stored as units consisting of bi- and trigrams. In a similar vein, Servan-Schreiber and Anderson (1990) proposed that the representation can be captured by "chunks" of varying sizes that emerge from a processing of the patterns of covariations among the letters. The chunking operations in Servan-Schreiber and Anderson's model involves higher-order representations than those suggested by Dulany et al. and Perruchet and Pacteau, but all three approaches share the essential feature that the mental representations assumed to be established in experiments on implicit learning is one intimately tied to the physical form of the stimuli.

When called upon to apply this fragmentary knowledge base, as when making well-formedness judgments of novel stimuli, the presumption is that subjects compare their fragmentary memorial representations with the "chunks" or local groups of letters present in the test stimuli and judge well-formedness by imposing a criterion based on the degree to which these codings match. These models share advantages and liabilities with the exemplar-based distributive models. Like them, their primary advantage is that they avoid the problems associated with establishing deep, abstract codes. However, they, too, must ultimately deal with the problem of the distinctiveness of memory and with the still unsolved problem of determining criteria for similarity. Also, since fragmentary models are, like exemplar-based models, tightly linked with the physical form of the input stimuli, they have difficulty in handling situations in which knowledge is transferred across stimulus forms. Nevertheless, there is a good bit of evidence that under the appropriate circumstances such fragmentary representations may be set up.

In addition to these clearly fragmentary approaches, there are two other models of implicit learning that share certain features and hence belong in this classification. These two differ from most of the above approaches in that they have been instantiated as formal computer-based models. The first is a model based on the Holland-type production systems (Holland et al., 1986) developed by Mathews and his co-workers (Druhan & Mathews, 1989; Mathews, 1991; Roussel, Mathews, & Druhan, 1990). It represents tacit knowledge using a classifier system along with a forgetting algorithm. In this model, which is dubbed THYIOS, memories are comprised of "chunks" of varying sizes based on estimates of optimal validity of each "chunk-size" given the task at hand. It is worth noting here that although the memorial content in this model is fragmentary, the approach that Mathews and his colleagues takes also emphasizes abstract features of complex displays. For example, in Mathews, Buss, Stanley, Blanchard-Fields, Cho, and Druhan (1989) it was noted that subjects can detect patterns of symmetry across halves of letter strings (e.g., PTTT.CXXX or VVXX.SSTT) and use these to make well-formedness judgments. Hence, Mathews's model should be viewed as one that has both abstractive as well as fragmentary elements.

The other model is Cleeremans's parallel distributed-processing model (Cleeremans & McClelland, 1991; Cleeremans et al., in press). This model is based on an architecture first introduced by Elman (1990) that utilizes a simple recurrent network (SRN). In Cleeremans et al.'s (in press) formalization, the SRN uses a three-layer back-propagation network to set up a representation of the complex sequential stimuli to which it has been exposed. The network progressively comes to capture the underlying structure of the input by closely approximating the conditional covariations among each element in the stimulus display and the possible successors for that element. Hence, it is capable of learning structured displays, such as those generated by simple artificial grammars, and of establishing a knowledge base represented as a set of conditional probabilities of all the possible successors of each element (letter) in the sequences. Such a representation is not fragmentary in quite the same way as Dulany's or Perruchet's characterizations, but it is fragmentary in that it is based on establishing connections that reflect the contingencies in the stimulus display on which it trained.

The evidence

All three of these general approaches to the representation issue have, as pointed out, particular benefits and liabilities, and these have been paraded by defenders and critics of each. However, as will become clear, there are reasons for believing that each point of view represents an idealized extreme not likely to be found in any ordinary data base. Although it should be clear that I feel that the weight of evidence supports the existence of abstract representations, it would certainly be wrong to suggest that only abstract representations emerge from implicit learning operations.

What is somewhat surprising, given the vigor of the disputes that have emerged in the study of implicit learning over the issue of memorial representation and the obvious interest in the representation problem generally,[4] is that there have been surprisingly few experimental studies dealing with this problem. This small literature base is due, at least in part, to the fact that the problem of mental representation does not lend itself easily to experimental manipulation. Of the studies that have been carried out, a few have attempted to deal with the issue by manipulating the manner in which learning takes place in order to discern its impact on representation. Experiments by Brooks (1978), McAndrews and Moscovitch (1985), Perruchet and Pacteau (1990), and Reber and Allen (1978) fall into this category. Others have used the transfer paradigm, a procedure that, in principle, should yield the most direct evidence. Papers by Brooks and Vokey (1991), Howard and Ballas (1982), Mathews, Buss, Stanley, Blanchard-Fields, Cho, and Druhan (1989), and Reber (1969) used this approach. In the review of this literature that follows, it will become clear that the evidence is, at best, ambiguous in providing support for any one or another of the views.

One early study that sought evidence for a particular form of mental representation of tacit knowledge was carried out by Brooks (1978). It used the artificial-grammar learning format in which subjects were given examples of letter strings generated by two separate AGs. Each string was presented to the subjects as the initial part of a paired associate (PA) with the name of either an animal or a city serving as the other part of the pair. The two AGs were, however, not linked with the animal-city distinction, which was, in fact, orthogonal to the two AGs. The animal-city distinction was merely a cover for the more subtle link: one AG generated strings associated with Old World animals and cities and the other AG generated strings associated with New World animals and cities. Brooks reported

[4] For example, Yoerg and Kamil (1991) recently glossed all of cognitive psychology as having but two domains, the study of the nature of processing and the study of the nature of representation.

that during a well-formedness task, subjects were able to successfully classify novel strings as either examples of Old World stimuli, New World stimuli, or stimuli that were neither, these, of course, being strings that were nongrammatical by both artificial grammars. Brooks argued that in order for subjects to carry out this classification task they must have been using an analogical process comparing each test string with the instantiated memorial representation that was set up during the initial learning. He argued that establishing two abstract representations corresponding to the two AGs would have been unlikely in the extreme to have been formed under these learning conditions.

However, Reber and Allen (1978) were able to show that the establishment of this kind of instantiated, exemplar-based memory was a consequence of the use of the PA technique during learning. They were able to replicate Brooks's findings when subjects learned using the PA procedure, even without the orthogonalized use of two AGs and the elaborate cover story. Learning in this condition consisted simply of presenting subjects with a set of strings generated by a single AG with each string associated with the name of a North American city. When subjects made well-formedness judgments after this procedure (i.e., they were asked whether or not each novel string could serve as the associate for another city) they showed clear tendencies to make decisions based on drawing similarity comparisons with instances from the learning set.

However, when the same subjects learned a different AG using a simple observation technique in which they rapidly perused a large number of exemplars, Reber and Allen found strong evidence of abstract representations of the structure of the grammar. Moreover, in a two-year follow-up there was still evidence of these separate representations (Allen & Reber, 1980). Since this study used each subject as his or her own control, it seems clear that the form of knowledge representation that gets established can be modified rather dramatically by factors as simple as the procedures used during learning.

McAndrews and Moscovitch (1985) reinforced this general line of argument by showing out that the number of exemplars of the grammar used during learning has an effect. If relatively few exemplary strings are used, subjects do not have sufficient information to permit induction of the underlying patterns and tend to set up memories based on the limited concrete cases they have seen. When larger displays are used, the tendency is to form abstract representations that reflect the patterns manifested in the stimuli.

Perruchet and Pacteau (1990) presented evidence that they took as supporting, if only indirectly, the fragmentary view. They ran two groups of subjects: Group 1 subjects learned using a variation on the observation technique in which the subjects were allowed 10 minutes to peruse a list of 20 strings generated by an artificial grammar; Group 2 subjects were allowed the same amount of time but were given a list consisting only of the bigrams (controlled for frequency) contained in Group 1's list. On a subsequent well-formedness task, both groups were able to discriminate grammatical from nongrammatical strings at greater than chance levels, a result Perruchet and Pacteau took as evidence for the establishment of a fragmentary representation. While such a result does suggest that fragmentary memories can be used to make such well-formedness judgments, the evidence is far from compelling. For one thing, the subjects that perused the full items performed significantly better than those who saw only the bigrams, which suggests that these subjects had indeed established representations that went beyond simple fragments. For another, the fact that some subjects can reliably make judgments using only fragmentary knowledge does not imply that this is the default process when the full displays are available.

The picture that emerges from these studies is hardly singular. If anything, it suggests that the human subjects in these experiments are rather flexible and seem capable of establishing different memorial representations under different acquisition conditions, even when presented with the exact stimulus inputs as they were in the Reber and Allen (1978) experiment. The outcomes from the studies using the transfer paradigm are a little less ambiguous in what they can tell us about representations, but here, too, there are problems. In principle, transfer studies would seem to give us the best chance of providing a data base that will speak to the representation issue. They have the advantage that the conditions under which learning and testing are carried out can be more explicitly controlled, which should provide insights into the nature of the subjects' memorial representations.

However, I know of only five studies that are relevant here (Brooks & Vokey, 1991; Howard & Ballas, 1982; Manza & Reber, 1992; Mathews, Buss, Stanley, Blanchard-Fields, Cho, & Druhan, 1989; and Reber, 1969) and all have used the standard artificial-grammar learning task. These five, like the preceding, consist of initially presenting subjects with stimulus items that are rule governed (they conform to the structure of an AG) and then evaluating what has been learned under a variety of conditions in which the stimuli are modified in some systematic fashion. The transfer element is, of course, introduced by these modifications, most commonly by changing the physical instantiation of the stimuli at some point in the experiment and observing the impact that this change has on performance relative to control conditions where no such change occurs.

The first study employing transfer used a simple memory procedure (Reber, 1969). Subjects were asked to memorize sets of letter strings generated by an AG. The experiment was a demanding one in that the strings were given to subjects in sets of four strings each. In experiments like this subjects typically show continued improvement in memory performance across sets as they gradually learn to exploit the structure embodied in the strings. Halfway through the procedure subjects took a short break. When they returned the stimuli were, without warning, modified. For one group, the AG that generated the letter strings was changed so that the same letters now appeared but in strings with a very different underlying structure; for a second group, the letter set used to instantiate the AG was changed but the underlying grammar was not; for a third, both the AG and the letter set were changed. A control group continued to work with unchanged strings.

The results were quite striking. Those groups of subjects who worked with strings generated by the same grammar showed positive transfer in the sense that their capacity to memorize strings was unaffected. This result was quite striking because there were no significant differences between those subjects who had the letter set changed and the control subjects. On the other hand, performance of those subjects who worked with strings generated by the new AG suffered dramatically. Basically, all subjects who continued to work with strings that had the same underlying structure continued to perform at a high level after the switch took place; all subjects who had the underlying rule system altered showed negative transfer in their ability to memorize letter strings.

On the face of it, these data certainly appear to lend support to the abstractive view. However, this study suffers from some shortcomings. For one, no attempt was made to see if subjects could generalize their knowledge to different kinds of tasks; transfer here was a continuation of the initial condition, subjects were still memorizing letter strings. For another, the strongest test for the capacity to transfer knowledge of underlying structure, a "scrambled" condition in which the same letters were used but in different locations, was not run. And last, the initial memorization phase of the study was so demanding that subjects had begun to build up proactive interference (PI) and to tail off in their performance by the time they worked with the last two sets of strings. Hence, after the break, all subjects showed some measure of improvement through the simple action of release of PI.

The first attempt to examine the possibility of transfer across sensory domains was carried out by Howard and Ballas (1982). They used a set of real-world environmental sounds (e.g., squeak, drip, hiss, clang, and flush) to instantiate an artificial grammar. During an observation-style learning phase, the stimuli were presented either visually (the written words) or auditorily (the sounds represented by the words). During the well-formedness testing phase, half of the subjects who had received visual stimuli during learning were tested with auditory stimuli. The performance of this group on the well-formedness task did not differ significantly from subjects who were presented with stimuli from the same sensory modality (visual or auditory) during the two phases, implying cross-modality transfer. There was no test for auditory to visual transfer in this study. This was an unfortunate lapse since recent work on implicit memory using a cross-modality priming task has suggested that such transfer may be limited to explicitly coded representations (Schacter & Graf, 1989) and, when it occurs at all, is from the visual to the auditory.

In a recent series of experiments, Manza and Reber (1992) have succeeded in finding evidence of transfer that occurs across both stimulus form and modality. In one experiment, subjects learned strings from an AG that was instantiated using one letter set during learning and a second letter set during the well-formedness test. Subjects were able to discriminate well-formedness at rates well above chance, although their performance was below that of control subjects who worked with the same letter set on both tasks. This result suggests that cross-modality transfer takes place but that the transfer is not complete.

In a second experiment the transfer was across stimulus modalities, the learning strings were presented auditorily to half the subjects and visually to the other half. On the well-formedness task, half of the subjects in each condition had to make decisions with strings presented either in the same modality as during learning or in the other modality, yielding a total of four groups. Under these conditions virtually complete transfer was found; all groups were able to discriminate well- from ill-formed strings with equal facility. Experimental groups were indistinguishable from control groups and no differences were found between the auditory and visual modalities. It should be noted that this complete transfer, while certainly not the norm in this area of study, is not terribly surprising since there are reasons to suspect that subjects internally rehearse the letter strings when they are presented visually and attempt to visualize them when they are presented auditorily.

Manza and Reber's third experiment looked for transfer across both stimulus form and modality. Here the visual stimuli used during learning were sequences of stars that flashed in specific locations on a computer monitor; the auditory stimuli were sequences of tones of varying pitches played through the computer. A fully orthogonalized transfer test was used so that all possible shifts from stimulus form and stimulus modality were counterbalanced. There was one additional instruction given to the cross-modality transfer subjects: prior to the well-formedness task, they were told that the lowest pitch tones corresponded to the left-most locations and the highest to the right-most. Subjects could make well-formedness judgments well above chance under all conditions, showing evidence of transfer across both stimulus form and modality. The transfer, however, was far from complete. As in the first experiment, the cross-modality transfer subjects showed poorer performance than same-modality subjects and, overall, those working with the visual displays performed slightly better than those working with auditory displays.

Taken as a whole, Manza and Reber's series of experiments provides some fairly strong evidence that transfer of knowledge can take place in implicit learning situations. It also provides evidence that suggests that such transfer can take place across both the physical form of the stimuli as well as the sensory modality in which they are presented. These findings suggest that the memorial instantiations that subjects establish are abstract and not tied to any particular instantiation. However, the failure for transfer to be complete under the more demanding conditions, such as those where the physical form of the stimulus is altered, implies that some knowledge is likely tied to particular stimulus instantiations.

Mathews, Buss, Stanley, Blanchard-Fields, Cho, and Druhan (1989) carried out what is probably the most heroic transfer study on record in which the effects of changing the letter set used to instantiate an artificial grammar were examined over a four-week period. In this study, each subject participated in four sessions, each one week apart. In each session, subjects were initially given a list of items generated by an AG and asked to learn them. All subjects were then tested using a multiple-choice-type task in which sets of five strings were presented, with each set consisting of one grammatical string and four nongrammatical strings with varying degrees of violation of the rules of the grammar. The subject's task was to pick the most grammatical string. For the experimental subjects, each week a different letter set was used to instantiate the grammar; control subjects worked with the same letter sets throughout.[5] Mathews, Buss et al. reported clear evidence of positive transfer in that the experimental subjects continued to perform well without any additional feedback trials. In fact, they reported that on a few occasions subjects actually improved in their performance when the letter set was changed.

Mathews's view on these and other data is that they support the general notion that tacit knowledge may be represented abstractly (Mathews, 1991, 1992). However, he has argued that it would be a mistake to conclude either that all such knowledge is abstract or that all learning in these kinds of studies takes place independent of consciousness. Mathews takes what he calls a "synergistic" view, in which implicit and explicit processes cooperate and interact with each other to establish mental representations that are largely abstract but with some instantiated features to them. From the functionalist's stance, there are reasons for agreeing with this position.

The only other examination of transfer of implicitly acquired knowledge of which I am aware is a study by Brooks and Vokey (1991). This experiment is the only one that controls the degree of both physical similarity and grammaticality between the letter strings used during learning and testing. During learning, subjects observed a set of letter strings generated by an AG using a single letter set. The well-formedness test consisted of a set of carefully constructed items that differed from those used during learning in the following specific ways. For each item used during the learning phase, there were four corresponding items used during the well-formedness testing phase: (1) one that was grammatical and "close" to its corresponding learning item, differing from it by a single letter; (2) one that was grammatical but "far" from its associated learning item, differing by several letters, (3) one that was nongrammatical but "close," and (4) one that was nongrammatical but "far." The transfer element was introduced by instantiating half of the items in the well-formedness test using a new set of letters, while half used the same set used during learning.

Brooks and Vokey argued that if subjects emerged from learning with a "pure" instantiated, exemplar-based memory, they would tend to make well-formedness judgments based on physical "distance" from the learning items. Such subjects would not reliably distinguish an item's grammaticality and would not show transfer to the new letter set. On the other hand, if subjects had a "pure" abstract representation, they would select items based on their grammaticality independent of physical distance and would perform well on the changed letter set. In fact, both patterns of decision making were evident in the data, with the "distance" factor producing a modestly larger effect than the "grammaticality" factor. Brooks and Vokey argued for a process based on what they called "abstract or relational analogies" based on an instantiated memory

[5] This study had a variety of other factors that were being examined in addition to transfer. Several of these have been discussed elsewhere (e.g., Chapter 2); here I am only concerned with the transfer data.

that has the capacity to detect relational components that extend beyond the simple physical form of the input stimuli. What is interesting about this study is that while it provides clear evidence for transfer of complex knowledge across stimulus domains, which suggests an abstractive process, it also supports the existence of a distributive process in which specific item analogies are an important contributor.

However, the picture blurs somewhat when other implicit learning tasks are considered. The only other experiments that explored the possibility of transfer of implicitly acquired knowledge used the Berry and Broadbent (1984) process control task, and both of these failed to find evidence of transfer. The standard process control task has two forms, an impersonal one in which subjects attempt to control a simulated manufacturing or production plant by adjusting factors like the size of the work force and a personal one in which subjects attempt to control the emotional tone of a "computer person" by modifying the manner in which they react to it. Berry and Broadbent (1988) found that subjects did not transfer knowledge gained while working under one form to the other, even though both forms used identical rule systems. Similarly, Squire and Frambach (1990) found no evidence of transfer between these two tasks, neither in their normal subject population nor in the amnesics that were run in this study.

At this juncture there is no obvious reason why transfer of implicit knowledge should be regularly observed in the artificial grammar-learning studies but not in the process control tasks. Lack of transfer in these process control tasks implies that highly specific representations have been established; however, other findings from these studies, such as Berry and Broadbent's (1988) finding that explicit instructions diminished performance on these tasks, are consistent with the results from experiments using artificial grammars. I hope further work will reveal what features in each task are responsible for this lack of consistency.

In summary, the evidence in support of one or another theory of representation in implicit learning is weak, at best. The several studies that manipulated the conditions under which learning occurs seem to suggest that memorial representations are plastic and are adjusted to suit particular demand characteristics. Those studies using the transfer paradigm have yielded results that suggest that an abstract representation is established, although Brooks and Vokey's (1991) findings suggested that instantiated representations are likely playing a large role. We are stuck here in one of those classic muddles in psychology. The organism, and its attendant processes, is apparently so flexible that it avails itself of a variety of different processing modes, each sensitive to the constraints of the task and the demands of particular settings. The question that remains unanswered is what would be the default mode, if there were such a thing? That is, "all things being equal" what is the mode of functioning that we would anticipate finding? There are no clear answers, only further speculation. So. . . .

Allowing myself some room to go beyond the data, I would like to suggest that there is a reason why each point or view manages to find itself supported by at least one circumscribed data base. It is that under the appropriate circumstances each of these three kinds of representations is going to emerge. As I argued in Chapter 2, the subjects whose cognitive capacities generate the data for our science are conspecifics of ours and are possessed of a familiar brand of cognitive flexibility. To date we have not looked sufficiently closely at the role of these various experimental settings we have been using and have not carefully examined their attendant demand characteristics and the impact they have on the cognitive processes of our subjects. Recall how easy it was for Reber and Allen (1978) to shift the same subjects from an instantiated to an abstract representation simply by altering the conditions under which learning took place—and recall that those manipulations were so robust that the cognitive residue could still be observed two years later (Allen & Reber, 1980). Recall also that McAndrews and Moscovitch (1985) were able to produce a similar shift in the cognitive processes that subjects engage simply by increasing the number of exemplars they were required to learn. Indeed, following up on this line of thought, it seems fairly obvious that as the number of exemplars of a particular structure or the number of episodes of a particular kind increases, the advantages that accrue to an instantiated memory become less and less obvious. Memory overload is still most easily handled by establishing cross-exemplar or cross-episode encodings based on the abstract patterns of covariations among the stimulus elements in the display.

But let's keep pushing this line of reasoning. A quick scan of the four studies yielding evidence of exemplar-based or fragmentary mental representations indicates that each had methodological features that would encourage subjects to set up exactly such a memory space. Specifically (1) in the first study that produced evidence to support the establishment of an exemplar-based representation, Brooks (1978) used the paired-associate technique during learning; (2) in a follow-up study, Brooks and Vokey (1991) used a relatively small sample of exemplars presented to subjects on a single piece of paper that they were free to peruse; (3) Perruchet and Pacteau (1990) employed a similar list-scanning procedure in one condition and in the other gave subjects a list of bigrams, which they were similarly permitted to peruse; (4) Dulany et al. (1984) used a more traditional learning procedure, but the manner in which he tested his subjects strongly encouraged the use of fragmentary codings.

A parallel argument can, of course, be made for each of the studies that found evidence favoring the establishment of abstract representations. They typically used many exemplars presented either in a difficult memory task (Reber, 1969; Reber & Allen, 1978) or asked subjects to observe large numbers of stimuli under conditions of rapid presentation (Reber & Allen, 1978; Reber et al., 1980). The latter condition is virtually guaranteed to work against the establishment of any exemplar-based memories since subjects are bombarded with a large number of similar stimuli so rapidly that they have little or no opportunity to engage in specific encoding of individual items.

Finally, there are reasons for suspecting that other simple manipulations can encourage the establishment of abstract representations. For example, in a recent unpublished study (Manza, 1992) it was found that the more general the initial learning, the more likely the subjects will induce abstract representations independent of specific instantiations. Using a standard AG learning experiment, two groups of subjects were trained by asking them to reproduce each of a set of 20 different letter strings generated by the grammar. For Group 1 subjects, all of the strings were instantiated using a single letter set; Group 2 subjects learned syntactically equivalent strings, but half of them were instantiated using a different letter set than the other half. All subjects were then tested on a well-formedness task using strings generated with the letter set or sets that they experienced during learning ("control items") as well as strings generated using new letter sets ("transfer items").

The results revealed an interaction between learning condition and the nature of the test items. Specifically, Group 2 subjects were better at judging the grammaticality of the transfer items than those in Group 1, but Group 1 subjects were better on the control items. The suggestion is that the degree to which subjects establish abstract representations is likely to be governed, in some measure, by the initial acquisition conditions. When learning consists of exposure to items with a single instantiation, subjects tend to "hitch their fortunes" to that single representation; when learning consists of several instantiations, they tend to form abstract representations that capture the general structure of the displays independent of any particular instantiation. The interesting aspect of these different modes of representation, particularly from the functionalist's stance, is that it is quite clear that each mode of representation has its advantages and disadvantages. This study, of course, needs to be replicated and extended to larger stimulus displays and different numbers of instantiations before we can be fully confident in this interpretation.

In summary, it seems clear that the only alternative still left standing is the functionalist perspective that I have been arguing for all along. It seems wrong to me to argue that representations must be of one kind or cannot be of another kind. Does implicit learning yield an abstract, tacit knowledge base? Yes, of course it does. Must implicit learning yield an abstract, tacit knowledge base? No, of course not. What is the default mode? Here we still do not know, although to be honest, I still have my intuitions. I suspect that when the displays are complex and the number of instances or episodes is large, the default representational mode is an abstract one. However, this is, as we like to say, an empirical matter and will be decided one way or the other with further research.

What I would like to see here is some kind of theoretical model that can handle a real-world situation that I observed recently. I was in an art museum and found myself quite overwhelmed by a large brooding and dark work by a (to me) unknown eighteenth-century artist. In the gallery with me were a half dozen or so others including a young couple, both of whom were also rather taken with this powerful piece. After a few minutes the young woman turned to her friend and mused, "You know, if Beethoven had been a painter that is the sort of thing he would have done in his later years." Virtually everyone in the room who overheard this comment nodded assent at her interpretation. It was immediately intelligible to all of us and we just as immediately recognized the legitimacy of it.

This is a wonderful example of transfer of abstract structure across stimulus domains. The common emotive components of late Beethoven composition and this painting were apparent to anyone familiar with them both. Just what shared elements gave rise to this sense of deep commonality are as unknown to me now as they were then. But all of us there were acutely cognizant that the transfer had been effected and that it was not dependent on specific fragmentary elements that two displays had in common nor on analogical judgments based on comparisons between instantiations. The transfer that occurred required that each of us there in that gallery have an abstract representation of the sound, texture, and emotive qualities of the later Beethoven works on one hand and of the painting on the other and the capacity to detect the structural isomorphisms between them. We are making progress on understanding the problem of mental representation, but we have a way to go yet.

On consciousness

In both Chapters 2 and 3 I allowed myself only a bit of room to discuss the topic of consciousness. In Chapter 2, I kept the discussion down to methodological issues; in Chapter 3, I permitted myself a little leeway to deal with some larger speculative issues that developed from the basic principles of evolutionary biology that were introduced. Here I would like to give myself more of an opportunity to extend that speculation.

Since one of the hallmarks of implicit learning, as I have characterized it, is that it is a process that takes place largely outside of consciousness, it seems reasonable to try to come to grips with what is meant by consciousness. After all if I am going to theorize that something lies outside of a set of mental boundaries, I am obligated to at least make a stab at articulating what those boundaries are. As will become apparent, this is no mean task. It is, moreover, a task confounded by a number of issues that were raised earlier. For example, at the beginning of Chapter 2 in a section entitled "The Polarity Fallacy," I argued that while it may important, for argument's sake, to maintain the capacity to differentiate between that which is conscious and that which is unconscious, in actuality it is unlikely in the extreme that there exists any sharp division between these two modes of cognitive functioning. Rather, it is both evolutionarily and epistemologically more sensible to think of a continuum of cognitive function that has elements that are utterly opaque to consciousness at one end and elements that are utterly transparent to consciousness at the other.

Also, there are the several methodological issues raised in both Chapters 1 and 2 under the general rubric "methodological indeterminacy." There, it will be recalled, I argued, following Erdelyi (1986), that independent of ontological issues concerning consciousness, the cognitive unconscious, and the underlying dimension that connects them, there are profound methodological problems associated with any attempt to assign particular functions to either end of this continuum. I tried to show in Chapter 2, I hope convincingly, that virtually every technique employed to determine the degree to which any given cognitive content is open to consciousness is compromised by methodological problems.

Both of these issues, ontological and methodological, imply that trying to articulate clearly just what is meant by consciousness is a futile gesture that will founder on these conceptual barriers. Nevertheless, there is this historically obvious, abiding interest in (if not absolute fascination with) consciousness, its functions, and its relationships with other mental processes and operations. Probably no topic in the history of philosophy and psychology has received more attention. Accordingly, I do feel obligated to at least put in my two cents.

First, I view consciousness, in whatever guise, from a functionalist, adaptationist point of view. Consciousness is a mental feature of a complex organism with the requisite neuroanatomical structures and organization. It is, moreover, a feature that has had a long but so far poorly understood evolutionary history, one I feel confident will eventually be found to display various phylogenetic and ontogenetic properties in accordance with basic principles of evolutionary biology. If you have the right kind of brain in the right kind of organism, that organism will have some sense of self and, in the more sophisticated versions, will have a sense of self that has self-referencing, monitoring, and reflective properties.

But a cautionary note is needed here: These emergentist considerations, no matter what role they may have played in various historically significant considerations of consciousness, do not necessarily invite the kind of dualist thinking that they often have and do not necessarily entail the Cartesian notion that consciousness is an all-or-none affair. The commonsense stance to take here is that of simple pragmatism. Consciousness is, like all other cognitive functions and processes, predicated upon and ultimately reducible to neuroanatomical properties. It must, in the end, be amenable to the same laws that describe other material systems. Having said that, I hasten to point out that these mechanistic, physicalistic considerations do not necessarily entail a movement toward the neurologically grounded reductionism favored by eliminative materialists (P. M. Churchland, 1986; P. S. Churchland, 1986). The recognition that cognitive processes are carried by particular neuroanatomical principles does not demand that discussion of cognitive processes must be in neurological terms. It is currently (and, indeed, may always be) most inconvenient to talk in neuroanatomical terms both because we don't have the language right now and because, I suspect, that even when we do we will find it to be far too cumbersome. It will, I believe, continue to be easier to gloss neuroanatomy with simple cognitive science terms; eliminative materialism is a naive point of view from the simple perspective of those actually doing science.

Now, given this little introduction, I would like to argue that is important to try to distinguish at least two different kinds of consciousness, kinds that, in accordance with the adaptationist's stance, seem to make a certain amount of evolutionary sense. It seems pretty clear that even the most primitive of organisms can be "credited" with some rudimentary forms of responsivity to the environment; even single-celled organisms react to noxious stimuli and make "deliberative" locomotions away from such. However, it is not clear that we want to credit a paramecium with consciousness when it withdraws from (notice, I did not say "avoids") a noxious stimulus. I, for one, am content to leave such organisms in a kind of mechanistic Cartesian state; their behavioral repertoires appear to be quite reflex-like and Lloyd Morgan's Canon should be our guide for understanding.

However, as we inch up the phylogenetic scale to a species that has attracted considerable interest lately, the giant sea slug, *Aplysia californica,* an intriguing issue emerges. This creature, with precious few neurons and barely anything that can be generously called a brain, clearly has the capacity for differential conditioning (Carew, Hawkins, & Kandel, 1983). Differential conditioning, while still well within the explanatory realm of Cartesian mechanism, suggests that something else may be emerging here; *Aplysia* is processing information and coding events. Whether or not this means that our modest slug is "aware" of anything is still surely doubtful—Lloyd Morgan is still likely with us—but the suggestion is that this beast has begun the process of distinguishing between "in here" and "out there" (see Bailey & Chen, 1991, who review much of the recent work on the neuroanatomy of learning and memory in *Aplysia*).

Let's take a large phyletic step here and move up to a much more sophisticated organism, the frog. Let me do a cursory gloss on a passage from James's *The Principles of Psychology* (1890, Vol. I, pp. 14–18). In this section he first presents to psychologists what has since become a standard overview of the relationship between various centers in the central nervous system and their corresponding behavioral functions. He outlines the relationship between various behaviors in a frog's repertoire and the underlying neuroanatomical structures that subserve them. A normal, intact frog has a fairly impressive range of behaviors, most of which appear to be rather tightly tied to particular stimulus events. Leg twitches accompany stroking along the frog's side, hopping follows a pinch of the hindquarters, tongue flickers are elicited by flying objects of appropriate size and flight patterns, and the like. In short, frogs do frog things and, moreover, they do them with a kind of poignant frogish demeanor. If one removes the higher brain centers of the frog, changes are, of course, noted. But the changes that are observed are not merely in the creature's behavioral repertoire. Indeed, virtually the entire range of behaviors are still present; strokes still elicit leg twitches, pinches hopping, and flies tongue extensions. What seems to be missing is not the raw behaviors, which are in large part mechanistic, reflexive, and wedded to particular lower brain structures, but the volitional elements, the (to use Skinner's term) *operant* aspects of action. The careful observer quickly notes that the frog, in so many words, seems to have lost its essential frogness.

I would like to suggest that what the decorticate frog has lost is its rudimentary consciousness. The sense that it had that allowed it to differentiate "me from thee"; to distinguish between self and not-self. This is not to credit frogs with anything like human consciousness for there is no reason to suspect that their primitive forms of selfhood have anything like the self-monitoring features that we associate with human consciousness. My point is simply that there are good reasons for assuming that a viable and significant form of consciousness is likely emergent in even fairly primitive amphibian forms. Let's call this form Consciousness I.

The kind of consciousness that lies at the heart of the arguments raised in the context of implicit learning theory appears, however, to be an epistemic entity of a rather different sort. The kind of consciousness we typically imagine when we think of our own sense of awareness is one that functions not merely to differentiate self from other but one that incorporates a large number of cognitive functions that allow us to modulate and refine the actions of self (see Baars, 1988, for an extended essay on the various possible functions of consciousness). This form of consciousness, which we can call Consciousness II, is the form that has attracted so much philosophical and psychological interest over the millennia. Consciousness II feels introspectively very different from the more primitive form, and there is precious little evidence that it is found in species other than the higher primates.

Consciousness II is the consciousness that is the focus of the various arguments put forward concerning those aspects of implicit learning that operate outside of consciousness. The point to note here is that in all of the experiments on implicit learning that we have run, we invariably find that our subjects are conscious of the fact that they have learned something. That is, they are aware of the existence of some cognitive change that has taken place during learning and of the fact that they know something they did not know before. This is manifested by the fact that when subjects are asked to give confidence ratings when they make decisions or solve problems in our experiments, these ratings always correlate with performance. Subjects know they know something; they simply do not know what it is they know. But this kind of consciousness is not really Consciousness II, it is something cognitively less well developed. It is, perhaps, more akin to the "cognitions" (using that term loosely) of James's frog with an intact cortex; it is a function that contains awareness but lacks the self-reflective, modulating functions that I would like to link with true consciousness—that is, Consciousness II. Knowing that one knows something and being able to either raise that knowledge to consciousness or overtly use that knowledge so that it has an impact on other behaviors is characteristic of knowledge that is represented tacitly. True Consciousness II is a much more sophisticated kind of cognitive function; true Consciousness II is characterized by both self-reflection and the capacity to use the knowledge derived from self-reflection to modulate other functions, to have a causal role to play in other cognitions.

When Descartes doubted all but that he thought, it was this latter form of conscious reflection on himself that represented the indubitable. When Locke argued that all that was mental was self-knowable, it was this conscious reflection that was the underpinning of mind. This theme is surprisingly coherent and surprisingly widely held. Descartes and Locke did not agree on much, but they did both accept the principle that understanding consciousness was the key to understanding mental life. For Descartes it meant that mind was rational, reflective, and known to its possessor. This reasoning consciousness, this capacity for thought and reflection, this ineffable awareness of self and the actions of self, this was for Descartes the essential feature of his humanity. It allowed for the clean dissociation between our species and all others, they being but brutes whose actions are governed by the mere mechanical and reflexive. The notion that one could acquire knowledge of a complex and abstract sort is unthinkable within a Cartesian framework. It would, among other heretical theses, imply that the gulf between *Homo sapiens* and the "beasts of the fields" were bridgeable.

For Locke, this primacy of consciousness meant that mind was equivalent to consciousness and that all thought was open for introspection. The limits of introspection were, prima facie, the limits of the mind. As Dennett (1987) put it, to a Lockean the notion of unconscious thought was "incoherent, self-contradictory nonsense." It should come as no surprise that Titchener, the last defender of the Structuralist dogma that introspection must be the prime vehicle for understanding the mind, was an unabashed Lockean. For Titchener, the notion that any interesting kind of mental event could occur and be opaque to conscious examination was near heresy.

Just as an aside here, it is worth noting that while most historians of psychoanalysis view Freud's preoccupation with sex and other "base" motives as the primary reason why his early writings produced so much outrage, there are good reasons for thinking that it was his dismissal of the traditionalist's view of consciousness that was even more disturbing to his intellectual contemporaries. Certainly philosophers were more distressed at the suggestion of unconscious mental processes than they were of the suggestion of infantile sexuality. The latter was, from either the Lockean or Cartesian tradition, ontologically possible, however disturbing. The notion of unconscious cognition, on the other hand, was viewed as a direct contradiction in terms. Freud's great epistemological sin was relegating rationality and consciousness to the back seat of a very large epistemological bus.

It should also be clear that my presumption of the existence of a modulating, self-reflecting consciousness in no way entails the Cartesian notion that consciousness (that is, Consciousness II) is an all-or-none affair, something that is either on or off. Following Dennett (1991), I would take a somewhat more Hegelian stance in which the notion of consciousness captures, like the notion of language, a complex set of functions that requires both a requisite neuroanatomy and the proper social relations to permit appropriate development. Thinking of language or consciousness as an all-or-nothing affair makes precious little sense from either the psychological or the biological point of view. Moreover, to reiterate two common themes, (1) the notion of a sharp demarcation between that which is consciousness and that which is not is nonsensical from both the psychophysical and the epistemological perspectives; and (2) it is futile to try to establish behavioral tests that will determine whether or not any given mental function or mental content is within the purview of consciousness; there are always going to be lots of fuzzy, borderline cases of material that is marginally conscious and lots of elusive instances of functions and processes that seem to slip in and out of the scope of personal awareness.

As a final bit of speculation here, I would like to suggest that the degree to which a particular process or its resultant mental content may fall within the scope of consciousness will be related to some extent to the evolutionary antiquity of that process and its content. Thus, if we look at the basic process of a biologically complex organism like ourselves, it is quite clear that we have virtually no access to the basic physiological functions that are dependent on phylogenetically old systems. Try as we may we cannot introspect into nor modulate liver functions and, as anyone who has ever suffered serious injury or had major surgery can attest, the management of pain is not one of the functions of consciousness. At the other extreme, when the more recently evolved functions of complex problem solving or language are considered, we find that consciousness has a rather effective modulating role to play. Indeed, we consciously think and plan what it is we wish to say, and we find most problem solving to be the product of introspectively controllable hypothesis-testing procedures. It seems likely that the rest of our functions could lay themselves out along a loosely constructed dimension of "introspectability" and that this dimension would show more than a passing correspondence with the evolutionary age of the systems that underlay each.

Be that as it may, the heart of the matter of consciousness, particularly when applied to implicit learning, lies not so much in specific claims about consciousness and the cognitive unconscious; virtually everyone is prepared to accept the proposition that much of what goes on mentally does so independent of awareness and reflection. The point where the sometimes angry disputes emerge is the claim that the induction routines observed in the standard studies yield a mental representation that is both unconscious *and abstract*. For example, Perruchet and Pacteau, in a recent rebuttal to several commentaries on their work (Mathews, 1990; Reber, 1990), made the telling statement: "We do not question human abstraction ability, no more than we question the existence of unconscious processes. What we do question is the joint possibility of unconscious abstraction" (Perruchet & Pacteau, 1991; p. 113).

This comment, I believe, represents the heart of the matter. It surely leads one to suspect that this issue is more than a simple empirical one to be determined by experimental explorations. Indeed, the vigor with which Perruchet and his co-workers have challenged virtually every finding of the work in implicit learning bespeaks of an agenda that is more than a simple attempt to keep "the other guys honest." In a similar vein, Dulany's often passionate critiques of the same literature suggest that more is at stake here than interpretation of a small data base. In his most recent article, Dulany (1991) attempts to show that not only are the claims of workers such as myself, Mathews, and Lewicki on implicit learning unsubstantiated, but the ancillary work on implicit memory and the neurophysiological studies on populations with various forms of psychological and neuroanatomical disorders reviewed in previous chapters is better dealt with from the point of view that these phenomena reflect ordinary conscious operations.

Now, when these alternative characterizations of implicit learning have been proposed, they have typically been put forward under an empirical umbrella. That is, the questions that have been raised concerning what I like to call the standard characterization of implicit processes have tended to focus on concerns with matters of methodology, data analysis, and data interpretation. This is, of course, the normal state of affairs in our field. However, the intensity with which these issues have been raised leads me to suspect that these are, at some rather deep level, disputes over the foundational characterizations of the cognitive processes presumed to underlie the phenomena associated with implicit induction. These issues, I suspect, are ones that derive from fundamental philosophical differences in the epistemological orientations that are taken to the study of cognitive processes.

These kinds of concerns are, perhaps not surprisingly, the ones that hold the greatest interest for me. Issues of methodology, data analysis, and theory are often subsidiary to those of philosophy; in a kind of Kuhnian way, the philosophical framework within which one works can have a major impact on the approach taken to basic problems in science. Indeed, the often unspoken and occasionally even unrecognized philosophical umbrella under which a scientist operates can function in much the same way as a Kuhnian paradigm. It can control subtly the kinds of problems one approaches for study and can have a considerable influence on the manner in which one reacts to the work of others. Behaviorists rarely explore the causal nature of mental states and often find themselves bewildered by contemporary cognitive explorations of mind; Cartesians aren't particularly concerned with nonrational action and tend to view statements of the arational nature of behavior as misrepresenting matters; Lockeans tend to operate under the principle that all mental processes are transparent to consciousness and regard the notion of an unconscious cognitive process as an oxymoron. Interesting, it's likely that those whose thinking reflects behaviorist, Cartesian, or Lockean qualities would neither classify nor recognize themselves as such. However, I suspect that the passion that lies behind the recent spate of criticisms of the work on implicit learning derives from the still lingering vestiges of both Lockean and Cartesian thought.

As I pointed out above, the one philosophic theme Locke and Descartes shared was that it was an essential feature of human mental life that true abstract cognitive functions were both open to introspection and causally directed by consciousness. The differences between Locke and Descartes were, of course, profound and ineluctable—Descartes focused on the rational components of thought and Locke concerned himself with the embellishment of associative structures; Descartes grounded his epistemology in the doctrine of innate ideas and Locke defended the new environmentalism. Yet the two shared a common ground—the notion that a true cognitive unconscious capable of the apprehension of the abstract nature of the complex aspects of the world about was an epistemological impossibility.

It is certainly worth entertaining the suspicion that some of the current attacks on the claim that implicit learning embodies unconscious abstraction are motivated by deep philosophical considerations about what is, in principle, cognitively possible. For myself, I would like to leave this issue in the hands of the experimentalists. I feel it is basically an empirical question whether unconscious abstraction can take place. From the point of view of evolutionary biology, I see no inherent reason why such cognitive process cannot be. In fact, in terms of representing patterns of covariations among the many elements of a rich, varied, and constantly changing environment, I feel that exactly just such a set of encoding functions ought to have had great adaptive value and likely emerged rather early phylogenetically.

Prediction and generation of events

General issues

In this section I want to deal with what might seem, at first, to be a silly question, namely, "If you know what's going to happen, does that mean that you know what's going to happen?" This problem has reemerged recently in the general context of the larger issue of differentiating between mental content that is explicit and conscious and that that is implicit and not easily available to consciousness. Before pursuing this line of thinking, I feel obligated to reissue the set of caveats that I discussed in several places earlier. Specifically, attempts to deal with this issue invariably stumble over two problems. The first is the problem of *experimental* or *methodological indeterminacy,* the basic point of which is that the degree to which any particular mental content or process can be determined to be within the scope of consciousness is dependent, to a greater extent than we generally appreciate, on the methodologies used. The second is the problem of the *polarity fallacy,* the basic point of which is that the conscious-unconscious division actually represents poles on a continuum and that there is not going to be any point that can be used to demarcate those mental contents that are conscious from those that lie below the limen of awareness.

Because of these two interrelated problems, one methodological and one ontological, there will never be any simple way to distinguish between processes that are conscious and explicit and those that are unconscious and implicit. We have been over this ground before (although, as Erdelyi notes in his most recent argument [1992], it still seems one that is resistant to understanding), but the reason for bringing it up again is a new slant on the issue that has been introduced, which gives every appearance of being likely to cause even more confusion. Put simply, the proposition currently being entertained in various quarters is that if it can be shown that an individual knows what is going to happen, this very act implies that he or she knows what is going to happen.

I appreciate that phrased this way it appears that this is an empty tautology. However, this issue will turn out to be more than just a play on words; it will turn on just what we mean by the word "know" in this proposition. To appreciate the problem, take two similar but potentially very different experimental procedures. In one, subjects are presented with sequences of stimulus events, say flashes of lights, that occur in any of several marked locations on a computer screen. The subjects' task is to press one of several buttons that correspond to the location of each target as rapidly as possible. In the typical experiment using this procedure, the stimulus sequence is dictated by a set of rules and the gradual diminution in RTs that is observed is taken as evidence of the gradual learning of those rules. This technique has been employed widely; see, for example, papers by Cleeremans and McClelland (1991); Cohen, Ivry, and Keale (1990); Howard, Mutter, and Howard (1992); Lewicki et al. (1987; 1988); and Nissen and Bullemer (1987). Generally, the results from such studies are taken as providing support for implicit learning in that the subjects' responses are rapid and automatic and take place independent of any overt thinking or planning about individual responses.

In the other procedure virtually the same circumstances pertain; subjects are presented with sequences of stimuli whose order is dictated by a complex set of rules, and they must make specific responses to each. The difference here is that rather than ask subjects to respond as quickly as possible to each event after it occurs, subjects are asked either to predict the location of each upcoming event before it occurs or to generate a sequence of representative events. This predict/generate procedure has also been widely used; see, for example, Cohen et al. (1990), Howard et al. (1992), and Nissen and Bullemer (1987). It was, of course, used in the past in the many studies on the probability-learning task that were discussed in detail in Chapter 2. Unlike the reaction-time experiments, in these latter experiments in which the subject is able to make such explicit, overt responses at rates significantly above chance, there is a tendency by some to use this behavior as evidence that the subject's epistemic contents concerning the display are held explicitly. After all, what could be more explicit or representative of a conscious process than a subject telling us exactly what is going to happen under such well-controlled circumstances?

This line of argument is quite seductive; for example, Cohen et al. (1990) take success on this generation as prima facie evidence of explicit knowledge of the sequence of events. Perruchet and Pacteau (1991), Green and Shanks (in press), and Dulany (1991) all embrace this line of thinking to some extent or another.

Unfortunately, this argument simply does not go through. It is one thing to know *what* will happen next; it is an entirely other thing to know *why* it will happen next. It seems manifestly clear to me that individuals may be quite capable of making "explicit" predictions about upcoming events based on "implicit" knowledge of the structure of the sequential displays or generating "overt" sequences of events that are based on "covert" representations of the underlying properties of the displays. The capacity to make these predictions or generate these sequences of stimuli carries no entailments about the underlying form of the knowledge base that enables them. We do this sort of thing, most obviously, in every sentence we speak or write. We generate words one hot upon the heels of its predecessor with virtually no explicit knowledge of the syntactic rules of the language that enable such behavior. And it is not simply the case that this is a process that was once conscious and has since become automatized and unconscious; we never knew the rules of syntax in the first place. The entire acquisition, retention, and generation sequence has been carried out, like all implicitly acquired systems, independent of both the process and the product of acquisition. As Ryle (1949) argued, there is a difference between "knowing that" and "knowing why."

Why the argument that explicit prediction or generation of events implies explicit coding has surfaced again is puzzling. In the 1950s and 1960s scores of experiments were carried out using the probability-learning procedure that seemed to show convincingly that subjects could make accurate predictions of upcoming events without awareness. Much of this work was already reviewed in Chapter 2 in the subsection on probability learning. However, I would like to discuss here the results of a new series of studies that should help to make this point even more clearly. These studies, which are based on a new experimental technique developed in our lab (Kushner, 1992; Kushner et al., 1991), will also allow us to develop further some of the issues raised above concerning rules and knowledge representation.

Studies with a complex prediction task

In these studies subjects are seated at a computer monitor with three boxes set in an inverted triangle on the screen. The boxes light up, one at a time, in sequences of length n. The subjects observe a sequence of n flashes and are then required to predict the location of the $n+1$st event based on particular features of the preceding n events. This procedure is extremely flexible in that n can be varied widely and there is a wide range of rules that can be used to determine the location of the target event.

In the procedure that has been explored most thoroughly, n is set at 5 and the target event is determined by the locations of the second and fourth events in the sequence; the first, third, and fifth events are irrelevant. The rule used is biconditional as follows: If the second and fourth events occur in the same box, then Box 1 is correct; if they occur in a clockwise relationship to each other, then Box 2 is correct; if in a counterclockwise relationship, Box 3 is correct. In the course of a single experimental setting, subjects see all 243 possible sequences of the first five events exactly once, which controls for any adventitious biases that might be produced by randomization or quasi-randomization of events and, importantly, eliminates the possibility of the kinds of frequency artifacts that Perruchet et al. (1990) discovered had contaminated the Lewicki et al. (1988) experiments on learning structured sequences.

In the standard experiment with this technique, the subjects are run through 10 learning sessions totaling 2430 prediction trials, 4 sessions under a rule-change condition, and 4 sessions where there are no rules, the target event being determined at random. The rule-change condition, by the way, is most devious; sequences that predicted Box 1 now predict Box 2, those that determined Box 2 now yield Box 3, and so forth. The random condition was introduced primarily as a check on procedure to make sure that no untoward elements had leaked into the design.

Figure 4.1 presents the mean prediction accuracy of six subjects over the full 18 sessions of the experiment. The results are quite striking. Subjects begin

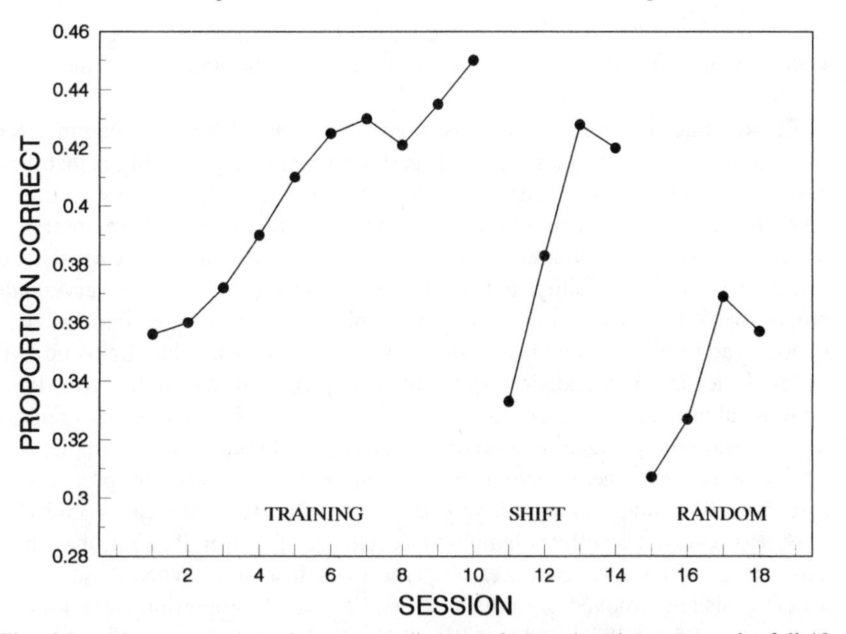

Fig. 4.1. Mean proportion of correct predictions of target locations across the full 18 sessions. Training sessions were run with the rule in effect; shift sessions with the changed rule; random sessions with no rule in effect. This figure is based on data from Michael Kushner's Ph.D. dissertation (Kushner, 1992) and is adapted from "Implicit detection of event interdependencies and a PDP model of the process" by M. Kushner, A. Cleeremans, and A. S. Reber, 1991, *Proceedings of the 13th Annual Conference of the Cognitive Science Society,* p. 217. Copyright 1991 by LEA Press.

predicting at chance levels and gradually increase the accuracy of their predictions until they reach average levels of about .45. It seems pretty clear that subjects are making predictions without overt knowledge of the rule used to generate the target events; if they knew this rule we would expect correct predictions to approximate 1.0. Interestingly, subjects differed very little from each other in the accuracy of their predictions. The poorest subject had an overall proportion correct (i.e., averaged across all 2430 trials) of .387, the best subject .444; on the final session of this phase of the study, subjects' proportion correct ranged from .407 to .498. Subjects had virtually no awareness of the rules in use, and in postexperimental debriefings none were able even to identify the second and fourth events as critical.

However, it is clear that subjects are learning a good bit about the underlying patterns of covariation in the sequences and the rule-shift transfer condition reinforces this conclusion. The first session here finds the subjects performing no better than chance (actually on the first 100 trials their performance was significantly below chance). With practice, however, they gradually improve and by the third session are averaging .42 correct. Note that no subject noticed the rule-change condition. Several subjects reported that they thought they had one or two "bad days" in the middle of the experiment, but no one noticed any specific shift in the rules that determined the locations of the target events.

The key question here is, of course, what is it that subjects are learning that enables them to achieve these levels of performance? One possibility is that they have learned just a few sequences and are getting these right virtually all of the time, thus dragging the overall proportion correct to the levels that are observed. Microanalyses of the patterns of prediction for individual subjects reveals some evidence for this possibility, but it turns out to account for only a very small proportion of the data. Some sequences are highly salient—for example, "continuous" and "near continuous" sequences in which the same box lights up five or four times in a row and single-alternation sequences in which the boxes light up in an alternating sequence such as 12121 or 32323. These sequences tended to be followed by higher-than-average correct predictions—for example, the "continuous" and "near continuous" sequences were followed by proportions correct of .611 and .501, respectively; the single-alternation sequences produced .439. However, these effects in no way account for the overall effect; they only make up 21 of the 243 sequences observed in each session. When these highly salient trials are removed from the analysis, the overall proportion correct drops from only .405 to .398, a value still well above chance.

It appears quite clear that subjects are picking up on the subtle patterns of covariations between events and using them to make accurate predictions and that they are doing this without recourse to overt or consciously held knowledge about the sequences. These results mirror strongly the findings from the earlier studies that Dick Millward and I carried out using the probability-learning paradigm during the 1960s and 1970s.

A simulation by a PDP machine

The data from these event prediction experiments have yet another intriguing feature: they lend themselves to exploration by at least one of the formal models of implicit learning that have been proposed in recent years. The model that we have examined in this context is the connectionist model developed by Cleeremans, Servan-Schreiber, and McClelland (1989; in press) that was discussed earlier. This model is based on a PDP machine and uses the simple recurrent network (SRN) architecture developed by Elman (1990).

To test the model, we (Kushner et al., 1991) trained the SRN by presenting it with the same sequences that our human subjects saw. During the presentation of the first five events in each sequence the network did not "learn"; it merely processed the sequence of events. When the fifth event was presented, the network was "turned on," and it was trained to activate the appropriate units. The results of this simulation are given in Figure 4.2, which compares the human data with the average proportion correct produced by the network.

It is clear that the model learns the regularities in the display during the initial phase of the experiment; in fact, it simulates the basic human data here quite

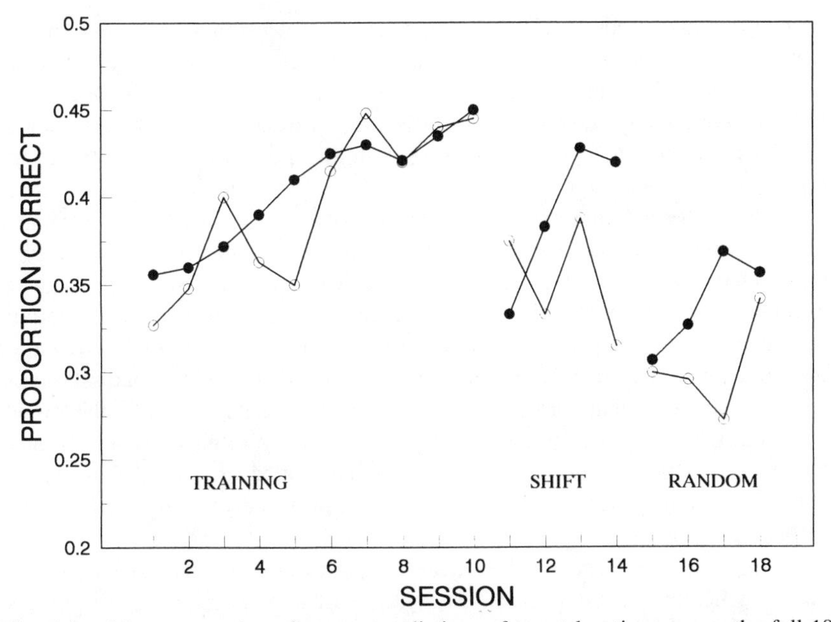

Fig. 4.2. Mean proportion of correct predictions of target locations across the full 18 sessions. Training sessions were run with the rule in effect; shift sessions with the changed rule; random sessions with no rule in effect. Filled circles give the actual data from the subjects; the open circles present the simulation data from the SRN. This figure is adapted from "Implicit detection of event interdependencies and a PDP model of the process" by M. Kushner, A. Cleeremans, and A. S. Reber, 1991, *Proceedings of the 13th Annual Conference of the Cognitive Science Society,* p. 218. Copyright 1991 by LEA Press.

nicely.[6] There are, however, other aspects of the network's behavior that do not correspond with the human data. For one, the network does not pick up on the salient patterns like the continuous- and single-alternation patterns; a concept like salience just doesn't get represented in a network that sees covariations as "raw" covariations. The model is also not very good at dealing with the rule-change condition; it has great difficulty in adjusting its performance to the new rule. This is important because it points to a shortcoming of virtually all current simulation models of implicit phenomena including those of Servan-Schreiber and Anderson (1991) and Jennings and Keele (1991), a relative inability to transfer performance to a different set of stimuli and responses.

Using a formal model of this kind to examine this task has helped to illuminate some of the issues of implicit learning. First, since the SRN has the capacity to

[6] The learning rate in the model is, of course, a free parameter. We experimented with several parameters; with some the network learned too quickly, with others, not at all. The simulated data displayed in Figure 4.2 are those that emerged from the parameter with the best overall fit with the human data.

encode complex, sequentially structured arrays in ways that are known to simulate human performance, it provides indirect evidence that analogous processes are being used by human subjects. For example, Cleeremans and McClelland (1991) have already shown that the SRN can learn the sequences generated by an artificial grammar in ways that closely mimics human learners. Second, the representations that are established by the model are completely opaque in the sense that information encoded by the network allows successful performance but is not readily decomposable into critical features. Indeed, it would take sophisticated analysis methods such as hierarchical clustering (see Cleeremans et al., 1989) to uncover the regularities embedded in the internal representations developed by the network. This characteristic of the representational system of the SRN (and, indeed, of PDP machines in general) provides a natural analog for the observation that human subjects are unable to make conscious the knowledge that is responsible for successful performance. Whether we really want to take such a parallel quite so literally is arguable (after all, the machine carries out no conscious processing), but it is surely a provocative interpretation of the SRN's behavior.

To summarize this section, it is clear that we cannot take the capacity to predict or generate upcoming events as evidence of an explicit cognitive process. When we put the findings from these new experiments together with the results from the several probability learning studies reviewed earlier it is clear that human subjects are able to make accurate predictions of future events with little or no conscious awareness of the complex knowledge base that they have acquired. In short, experiments that are based on prediction and generation behaviors can be used to explore implicit learning.

In fact, there are reasons for suspecting that experiments of this kind will turn out to be optimally designed for the examination of implicit learning. First, they are even more removed from real-world settings than the other procedures that have been widely employed like the artificial-grammar learning technique, Lewicki's person-perception procedure, and Broadbent's production-control task. As I argued in Chapter 1, if we hope to see implicit processes emerge in controlled laboratory settings, we need to establish conditions that are remote from real-world environments. As a corollary to this point, the prediction experiments have even less in the way of structure that is likely to be part of any subject's preexisting knowledge when they enter the laboratory. Second, these experiments can utilize materials that are presented rapidly, specifically so rapidly that subjects do not have the opportunity to encode them consciously. A feature of many other procedures is that the experiments have learning phases that invite some measure of conscious processing, for example, in the standard artificial-grammar learning task, subjects typically engage a variety of overt mnemonic strategies for memorizing the strings of letters. Third, this technique has the additional advantage that the kind of knowledge acquired in it is exceedingly difficult to articulate and, given the commonly held notion that there are links between consciousness and language, further supports the notion that this knowledge is held tacitly.

Nativism and empiricism

This is a difficult topic to write about. It is far more than a simple dispute over whether knowledge is given innately or provided by experience. These two terms have come to stand for clusters of issues that are complex and rich with philosophical, genetic, physiological, psychological, and sociopolitical argument. In this section I will try to fulfill the promise I made in Chapter 1 where I mentioned that the relationship between the examination of implicit learning and the nativism-empiricism debate would be explored and its entailments examined. Most of what I have to say will deal with the foundational issue of the origins of knowledge; some of the discussion, however, will hedge over into the associated sociopolitical issues. As I said, it is not an easy topic to write about.

I am going to use the terms "nativism" and "empiricism" as umbrella labels, each of which covers a variety of approaches to particular philosophical and psychological issues. Each is predicated on a core assumption concerning the origins of "critical elements" that permit of certain behaviors. In the standard classification, nativists assume that the critical epistemological elements are given a priori, that is, the individual is presumed to come equipped with the requisite knowledge to enable effective behavior. Empiricists, on the other hand, have traditionally downplayed the role of the a priori and assumed that knowledge is the result of particular environmental experiences. The term *nativism* was first used as a doctrine to mean that particular mental abilities—in particular, the capacities to perceive time and space—were inborn. In this form the doctrine defended the position that the mind's capacity for normal, three-dimensional perception existed independent of visual experience. This point of view was much debated, supported, and ridiculed in Locke's time and well into the nineteenth century, although it is no longer defended seriously by anyone (see James, 1890, Volume II, chapter XX; Morgan, 1977). In its contemporary, more moderate forms, nativism has come to represent those orientations that stress the genetic, inherited influences on behavior and thought over the acquired, experiential influences. The most vigorously defended contemporary form of nativism is, of course, that articulated by Noam Chomsky with respect to the ontogeny of language (see Chomsky, 1980, 1986, among many others). For reasons spelled out below, I will refer to this version of nativism as "content nativism." Because the Chomskyan orientation represents the prototypical nativist theory and the one most consistently defended these days, I will use it as a reference point in many of the issues discussed below.

On the other hand, the framework within which the study of implicit learning has taken place has focused on the role of the environment and regards it as playing a critical epistemological role. As I mentioned in Chapter 1, my orientation has been one that takes seriously Dewey's (1916) recommendation that the empiricist perspective operate as the default mode with nativist notions only being introduced when the data base forces them. Note that while this approach reflects many of the classic features of early empiricism, the pure empiricist position is no longer defensible. For reasons spelled out in the following pages,

I will use the term "process nativism" as a contemporary cover for the empiricist orientation.

The dispute over the degree to which various elements of thought and action are given a priori has been among psychology's most tenacious. The debate is hardly a child of this century. It hasn't tumbled to any substantive data base yet, and I hardly expect that it will in the near future. There are other factors driving the debate, and it is important to recognize them—deep issues of philosophy and ideology and passionately held beliefs about particular aspects of human nature. This is easily seen in the often powerful ideological and polemical writings of those who defend any of a variety of nativist-oriented lines of thought (e.g., Chomsky, 1977, especially chapters 1 and 4; Davis, 1986) and in the equally passionate works of those who have critiqued them (e.g., Lewontin, Rose, & Kamin, 1984). Of concern here are questions of essential preferences for freedom and independence, the interrelations between states and their citizens, the mutual cooperative arenas of science, government, and public policy, and so forth. In the end, the various approaches to the issue of the a priori can only be understood within this larger framework. However, for the purposes I have laid out for myself in this section, I will for the most part skip these ideological considerations and focus on those aspects of the debate that touch on the general issues raised by the examination of implicit learning and related topics.

Since psychology began slowly disentangling itself from philosophy in the mid-1800s we have agonized at length over this issue, with no clear winner. As points of view within the field and in the larger sociocultural context have waxed and waned, there have been varying amounts of enthusiasm for each position. In the middle decades of this century, during the heyday of behaviorism, empiricist thinking dominated and behavior was viewed primarily as the product of experience. Since the late 1960s a distinct shift has taken place, with nativism in any of a variety of guises becoming a powerful intellectual force in contemporary psychology. Its influences are felt in many areas beyond the study of cognition; they may be found in the examination of the roots of intelligence and intellectual functioning, in the search for genetic foundations for various disorders from alcoholism to schizophrenia, in those approaches to the study of personality and social psychology that have been guided by sociobiological principles, and of course, in the cognitive sciences. In what follows I want to raise a number of issues that are entailed by these apposable perspectives.

Varieties of nativism

As a starting place, it needs to be recalled that these days no one defends a "pure" empiricism in the sense of Locke's tabula rasa; it is uniformly recognized that some form of nativism must be true. It simply makes no sense to try to imagine the mind/brain as a homogeneous organ like a bowl of jello. Whether this compound entity is viewed from a perspective that focuses on the epistemological (i.e., the mind) or the neuroarchitectural (i.e., the brain), it needs to be conceptualized as having considerable structure, structure that is the end product of evolutionary processes that have winnowed the genetic possibilities down to those that have adaptive value.

Following terminology used elsewhere (Reber, 1973, 1987; Slobin, 1966; Winter & Reber, in press), the critical distinction is that between a *content*-specific nativism and a *process*-specific nativism. These labels were introduced by Slobin (1966) and neatly capture an important distinction. Content nativism assumes that the very content of mind, specifically the detailed content of encapsulated cognitive modules, is specified a priori. Process nativism, on the other hand, assumes that general processing systems that operate independent of specific modalities and of the input form of the input stimulus are given a priori. To take the two classic cases: Chomsky's nativism is content specific in the sense that specific mental content in the form of the deep structure of universal grammar is assumed to be laid down in the genes, whereas Piaget's nativism is process specific in that general cognitive processing systems that operate independent of specific cognitive or perceptual modules are assumed to be given. For those who are interested, a deeper sense of the distinction between these two orientations may be gleaned from the arguments made in the debate between Chomsky and Piaget and their respective supporters that was held some years ago (Piatelli-Palmarini, 1980). When cast in these terms, the contemporary version of empiricism within which implicit learning theory operates, is, as pointed out above, a form of process-specific nativism.

The tension between these two epistemological stances has implications for a variety of related issues in the cognitive sciences. The two that I want to focus on are the study of *learning* and the role of the *environment*. Content nativism necessarily downplays the role of learning, trading it in for a kind of biological unfolding process that is assumed to be dictated primarily by genetic factors. And, because content nativism places so little emphasis on learning, it provides little encouragement for the detailed examination of the impact of the environment. Process nativism, on the other hand, puts greater emphasis on both learning and the environment.

Learning

The basis for the current neglect of learning by nativists derives from Cartesian considerations about the nonlearnability of particular forms of knowledge. Historically, the inconceivability of acquisition has been a root stimulus for entertaining nativism. Descartes, in the first articulation of what we now call the "impoverished stimulus argument," could not comprehend how perfection could be a phenomenological reality when one is surrounded by the imperfect, so perfection became an innate idea. Reid, mocking the attempts of the British Empiricists to explicate spatial knowledge as wholly derived from sensation, separated perception as a distinct process and viewed it as being guided by innate structures. Kant, despairing over Humean skepticism and failing to comprehend how perceptual systems could organize and represent the complexities of the distal display, assumed that organizational tendencies that enriched and coded the environment were innately given. In more recent times, parallel arguments have been put forward by Chomsky (1980, 1986) with respect to language and language acquisition and Fodor (1975) with respect to concepts and concept learning. From this point of view, content-specific nativism is a position that would appear to be "forced" in that if there is no way that the environment to which one is exposed can provide the foundations for the requisite knowledge, then it must, perforce, have been given a priori.

However, there is another, perhaps less sympathetic gloss that can be put on this line of thought. It is not difficult to envision a more subtle and less lofty form of motivation operating here: simple frustration over the inability to develop a satisfactory theory of learning. This is seen clearly in the study of language and language acquisition. For many years now linguists have been unsuccessful in their attempts to explicate formally the underlying structure of natural language while every normal child possessed of only the most humble analytical skills apparently induces that structure without hardly working up a sweat. Unable to write a grammar for natural language, unable to conceive of the manner of its acquisition, and faced with countless examples of both its possession and induction, nativism became a most seductive doctrine. With this simple epistemic device, one is relieved of the burden of demonstrating how knowledge got there; one merely assumes that it was "there all along."

Empiricism, on the other hand, enjoys no such largesse, no simple escape from the problem of acquisition. As the early British practitioners became painfully aware, they had to do more than merely critique the foundations of nativist thought; they found themselves under an obligation to provide some coherent, at least in principle, mechanism(s) through which one could understand how knowledge acquisition occurred. If you assume that experience writes large upon the mind, then you are under a severe obligation to articulate some process or processes through which such writing tasks place. Hence, within the empiricist, process-oriented approach there will be, of necessity, a strong focus on learning and the processes of knowledge acquisition.

The role of the environment

In like fashion, the emphasis that is placed on the role of the environment is very different under these two forms of nativism. Content nativism views the environment as primarily a trigger of an unfolding process that is genetically specified. The environment is acknowledged to function to an extent to shape the development of knowledge, but it is not viewed as a significant contributor; the essential epistemic content of mind is not viewed as having been derived from the environment but from preexisting systems. Again, this is seen most clearly in Chomsky's approach to language where the emergence of a child's natural language is seen as a process in which the linguistic input to the child fixes parameters in the innate, underlying universal grammar. Empiricism or process nativism, on the other hand, places considerable emphasis on the environment. From the realist point of view of implicit-learning theory, the essential features of the environment are assumed to become mapped onto the memorial representational system. From this perspective a good research strategy is the one outlined by Mace's (1974) wonderful quip, "ask not what's inside your head but what your head's inside of." The thing that our heads are inside of is very complex; I tend to believe that it will, with more intense examination, turn out to be even more complex than we currently suspect. Those who have followed this advice have discovered that the stimulus display is quite a bit richer in information than earlier workers had believed (see, e.g., Shaw & Pittenger, 1977; Shaw & Turvey, 1981).

It will prove illustrative to pursue this point a bit within the context of the problem of language acquisition. Content nativists, following Chomsky's lead (see Pinker, 1989), generally regard the infant's linguistic environment as "impoverished" in the sense that the language corpus to which the child is exposed does not display the necessary linguistic variations that ultimately come to be part of the fluent speaker's underlying grammar. As noted above, this "impoverished stimulus" argument has a long history going back to Descartes, a philosopher for whom Chomsky has a special affection. Note, however, that this argument as it is typically put forward is predicated utterly on analyses of the linguistic corpus; virtually no account is taken of the social, interactive environment that supports the linguistic.

This strategy simply will not do; language does not take place in a void. Language is merely one (albeit large) part of a social, communicative, interpersonal, interactive system. Although it may be the most highly sophisticated medium for communicating things inside of one head to another head, it is not the only medium so used. Although it may well be the case that the raw linguistic corpus presented to the child is impoverished, this linguistic material is not the sum of the communicative messages the child receives. The purely linguistic environment is supported by a rich, pragmatic, performative, communicative, social environment that is, in all likelihood, not impoverished. The content-nativist position invites us to view the child as having an odd kind of tunnel vision. The child is viewed here as a social organism who, for unknown reasons, refuses to utilize contextual, gestural, and other paralinguistic information to buttress the meanings and intentions of the utterances that surround him or her. Since there has yet to have been the needed careful analysis of this linguistic supporting structure, I find content nativist claims based on the "impoverished stimulus" argument premature and unconvincing.

The presumption of content-specific nativism also tends to make us forget just how much learning actually takes place. Once again, an example from the Chomskyan-inspired approach to language will help to make this point. The speed with which the child's first, natural language is acquired is often cited as compelling evidence that the underlying structure of universal grammar must be inborn. The common argument is that language learning begins during the second year of life and is virtually complete by the sixth; a time frame that is generally regarded as too rapid to be the result of the operation of general learning processes.

However, there is actually very little reason to think that language is acquired rapidly, despite the many claims to the contrary. What is needed to make this point here is a little counting: There are 24 hours in each day, 365 days in the typical year. Let us recognize that for the typical infant the language-acquisition process begins, not during the second year, but at roughly six months of age, which is when there are clear elements of infant vocalization that are linguistically responsive, when the vocalizations of the normal infant become distinguishable from those of the congenitally deaf child.[7] Let us also recognize that the essential grammar-acquisition phase is not completed by age six. For the sake of argument here I will assume that it is complete by age eight—actually it's probably a good bit later than that, but this point is not really important. Now, let us allow our typical child an 8-hour night's sleep, a 2-hour nap, and another 4 hours out of range of communicative interaction, yielding roughly 10 hours each day during which language learning could be taking place. This estimate is probably an underestimate for the seven- or eight-year-old who is likely spending more hours per day in social interaction, but it is probably an overestimate for the very young child. No matter, it gives us an estimate with which to work. Simple multiplication shows an upper bound of 27,365 hours for potential linguistic and paralinguistic input to be analyzed. I grant that this total is probably only reached in highly verbal homes with educated parents where strong emphasis is put on verbal skills. But no matter; what we are really looking for here is some lower bound, some estimate of the minimum amount of time available for a child to engage the language-acquisition processes.

So, for the purposes of reaching some reasonable estimate let us take what I would guess to be a very conservative number: instead of 10 hours a day, let us try 3 hours a day. Here we still find a very large number, 8,214 hours of practice in one's native language. Cut the estimate to 1 hour a day and we still end up with 2,738.[8] That is a lot of time, a sobering amount of time. Indeed, it seems to me to be an extraordinary amount of time for a system that is supposed to have its content specified genetically and only to require the environment to act to trigger growth and fix parameters. For whatever it's worth, the U.S. Army Language school considers a mere 1,300 hours sufficient to attain fluency in a non-European language like Vietnamese (Burke, 1974).

Moreover, there are some interesting estimates that have been made with respect to the amount of time it takes to learn the structure of any rich, complex, abstract domain such as those that underlie natural sciences like physics, social sciences like psychology, or performing arts like music or dance. Numbers between 1000 and 5000 hours keep cropping up, no matter what the discipline (see Lindsay & Norman, 1977, p. 563). There is something very intriguing about the possibility that it takes roughly the same amount of practice to learn chemistry, chess, psychoanalysis, or the guitar as it does to learn one's native language. Perhaps these kinds of acquisition processes are guided, not by distinct modules

[7] Actually it is likely to begin even earlier; infants as young as 2 months can discriminate between speech items that differ from each other by a single distinctive feature such as /d/ versus /b/ (see Eimas, Miller, & Jusczyk, 1987, for a review).

[8] Notice that all of these estimates of experience still do not include the solo practice that Weir (1962) has documented in which children talk to themselves when alone at play and in bed. As she pointed out, this talk is no mere repetitive babbling of established verbal operants but systematic practice with structural forms in which alternate patterns of phrasing and word usage are generated and played with in a creative manner.

with specific structures, but by some set of general induction processes that operate in a wide variety of situations. Understand, I don't mean to try to "cash this check" here; we do not have anything like the kind of learning theory needed to capture these processes. But by assuming an innate, content-specific faculty of language that is represented by a distinct module, one is easily misled into thinking that the acquisition process is simple, rapid, and independent of other modes of acquisition. Moreover, one is also seduced into dividing up the domains of cognition and glossing over likely commonalities. Assuming a more general, sophisticated induction routine such as that implied by an empiricist orientation allows one to be more open to similarities in process as well as possible similarities in the nature of the environment within which each of these various cognitive capacities is acquired.

Some entailments of evolutionary biology

I want to raise two specific issues that become important when nativist doctrine is examined against an evolutionary backdrop—namely, *neoteny* and *mutation;* these issues involve features of human cognition and its emergence in the normal member of our species. *Neoteny* is the "retention of juvenile characters by adult descendants produced by retardation of somatic development" (Gould, 1977b); *mutations* are the saltatory changes in genetic material brought about by factors other than Mendelian recombinations.

NEOTENY. Although this issue is properly an intergenerational one, I want to deal with it here in its most obvious manifestation, the long childhood of members of our species. The several reasons for it emerged over the history of the species. For one, we are born relatively immature for the sake of the female, who has certain anatomical limitations on the size of the offspring that she can conveniently bear. It seems reasonable to assume that it was evolutionarily advantageous to bring forth a less completely developed neonate than to modify the anatomy of the female, for that would likely have reduced survival chances in terms of locomotion, fleeing danger, and so forth. Second, a lengthy period of relative helplessness in the face of a hostile environment was quite possibly an important element in the emergence of pair-bonding and the development of the family structure. Third, because of the importance of the social matrix for the protection of each member of the species, an extended period of infancy becomes critical for the development, refinement, and induction of a rich socialization system so that at maturity each member will be able to function in a socially effective manner. Fourth, the sheer amount of information needed for human adulthood is so great that the period of childhood has expanded to provide the requisite amount of time for its acquisition. This last factor is the one that has been emphasized most often by psychologists (e.g., Piaget) and social anthropologists (e.g., LeBarre), and it is the one that seems most relevant to the issues at hand.

There are two branches to the general issue of neoteny. One, the direct one, is how to explain the long childhood. The other, less direct but more relevant here, is how to explain the fact that some of the skills requisite for functioning seem to emerge relatively early in life and develop rather rapidly while others are delayed and take longer to reach a "steady state." In species where the ecological conditions demand it, we find early and rapid development of behavior and relatively short "childhoods." For example, many prey species like the ungulates are practically born running. A caribou calf runs with the herd within minutes of birth. Clearly locomotion is innate here and the most modest of environmental inputs is needed to trigger the behavior. Songbirds, on the other hand, play out their ontogeny a bit differently. The song is delayed and considerably more environmental input is needed to trigger it. Interestingly, the input can be degenerate to a certain extent, yet it will still be sufficient for a normal song to emerge at the proper time and with respectable rapidity of acquisition. On still another hand, marsupials are extraordinarily immature at birth but develop along rather tightly specified lines fairly rapidly thereafter. The point of all this, which is rather obvious and well-known, is that each species selects particular strategies geared to ecology.

How do we view our own species in this light? What nags at me here is why should we have this long childhood and why should our most poignantly human characteristics develop the way they do? What is a good way to conceptualize the manner of operation of those selection pressures of the past that have resulted in the creature we have become? Why, if language (to take again the example that is driving this discussion) is such an essential component of our cognitive repertoire, does it take so long to develop? If its essential underlying form is given a priori as the content nativists argue, why does it not develop much sooner and more rapidly than it does? Why does it seem to require the same amount of time and experience for acquisition as other complex abstract knowledge domains? Why don't we learn to speak like a horse learns to run or a bird to sing? I don't mean this to sound like a silly question, which on the surface it may seem to be. For example, our capacity for three-dimensional vision seems to develop early and rapidly. This is particularly interesting because, as pointed out above, the issue of vision is the one over which the nativist doctrine first emerged. Yet it has been some time now since researchers in vision seriously entertained strong nativist leanings; indeed, as a two-decades old quote from a learning researcher shows, eclecticism is the order of the day: "no psychologist today would espouse a simple nativistic or empiricist theory of perception . . . while the nature-nurture controversy is still a contemporary issue, psychologists in vision research tend to be less dogmatic than in the past" (Dodwell, 1975).

Why then the long childhood, and why does the "steady state" of mature language facility take so long to reach? The long childhood is an evolutionarily determined result of a large number of factors. The interplay between genetically controlled ontogeny, social and cultural development, the acquisition of language and its use in thought and other cognitive processes is extraordinarily rich and complex. In order to function effectively, the child must know a great deal about all of these elements; a child must know language to be able to communicate, but he or she must also have become inculcated with society's norms for proper action, have learned any number of specific roles, male/female, sibling, friend, and so forth. It seems to me unlikely in the extreme that hard-wired, content-specific systems could have evolved to satisfy this multifaceted system. A flexible, process nativism makes inherently more sense.

MUTATIONS. A useful heuristic in evolutionary biology is the analysis of mutations in particular forms and structures. If a precise phenotypic syndrome can be identified in a specific form, it can be taken as a critical datum for hypothesizing about the underlying genes that code for that form. Moreover, the degree of differentiation between those aspects of the phenotype that are affected and those that are not provides some insight into the extent to which the underlying genes are independent or mutually interdependent. Polydactylism, the presence of supernumerary fingers or toes is an example of a well-articulated syndrome, sharply delineated and functionally independent of other behaviors and structures. It is caused by a faulty gene that codes for the development of superfluous structures and has no other interactive manifestations in structure or function; sensorimotor function is not impaired and dexterity of the other fingers or toes is normal. Or take another classic case, the simple form of dichromacy where chromatic discrimination in the red and green regions of the spectrum is disturbed. Here, too, we find normal adaptive functioning in other aspects of color vision; the mutation is precise, local, and independent.

From the general characterization of content nativism, we would expect to find specificity of these kinds in cases where genetic mutations have compromised some aspect of the form or function of some element on the system. In the case of process nativism, we anticipate finding rather different patterns of disorders and dysfunctions. Here we would expect to find rather global and general disruptions of form and function since a mutation that interfered with the normal ontogeny of a general processing system would find itself expressed in a variety of processing domains.

The evidence tends to favor the process point of view, particularly when language functions are examined. There has never been to my knowledge an instance of a child who failed to show the capacity to display, for example, the specified subject condition or reciprocal expressions, or the *wh*-island constraint in their production or processing of language, who did not also suffer from an array of other linguistic and cognitive debilities. We do not observe cases where a child is missing a particular set of transformation or, interestingly, reveals a nonrepresentational production system that is fundamentally different from that in the immediate linguistic environment. When linguistic dysfunctions emerge, as they often do, they are accompanied by other problems, such as severe retardation in which all inductive processes are compromised and acquisition of cognitive functions are uniformly poor, or developmental language disorders in which the linguistic difficulties are accompanied by other, more general dysfunctions such as a deficit in responding to rapidly changing stimuli or in processing of sequentially ordered stimuli.

It seems fairly clear that factors such as neoteny and the patterns of mutations observed in various domains do not lend much encouragement to those who would argue for a content nativism with respect to language. Since language has been the standard bearer for this point of view, it is likely that close examination of other aspects of cognition would lead similarly to the suggestion that a more moderate, process-specific orientation captures the data more effectively. Note that one of my reasons for introducing these two issues from evolutionary biology was simply to point out that one of the failings of content nativism is that it makes little biological sense. It is far from obvious how the genetic foundations requisite for universal grammar should have evolved in the relatively short period in which the *Homo* line separated from the other primates. The existence of a rich biological foundation for language (Lenneberg, 1967) is unquestioned, but as many fail to appreciate, such a foundation does not entail the presumption of a *content-specific* form of linguistic nativism. It is perfectly reasonable to assume that we come equipped with a strong proclivity for language and other forms of communication and that we engage general induction routines that operate across modalities and across stimulus forms to induce the underlying structure of the stimulus environment within which we must learn to function. If such a picture turns out to capture the nature of language acquisition, it will provide considerable encouragement to those who argue for a process-specific form of nativism.

An overview from a functionalist perspective

It should be pretty clear what I am against in regard to the nativism issue. Let me end this section a bit on the upbeat with a few lines about what I am for. I favor, as I pointed out in Chapter 1, a rich, pragmatically guided interactionalism, a form of empiricism (process nativism) that goes by that much-abused label, *functionalism.* It makes precious little sense to me to try to conceptualize, let alone theorize coherently about, the nature of complex cognitive processes such as language or implicit learning "in isolation."

Functionalism carries with it, usually implicitly, the metaphysical principle articulated nowhere better than by Jacob Bronowski in his last substantial work (Bronowski, 1978). He was concerned about how to make a reasonable science, any science, when we must recognize to some extent or another that everything in nature affects everything else. For the psychologist this translates to the statement that there are conditions under which people will evidence one behavior pattern and conditions under which they will exhibit another. There is, I trust, no difficulty with this little truism. If we wish to understand what indeed our subjects actually do—and in the end we are all stuck with this litmus test even though we may differ in the manner in which we evaluate it—it seems trite that we won't get very far if we only look at the behavior and whatever hypothesized factors underlie it without recognizing the role of the conditions under which it emerged and the functions that it has.

This philosophic homily, of course, carries problems of a very real nature. Given Bronowski's metaphysical principle, there is no end to these considerations of role and function. What must be faced is the classical "boxing" problem. The inescapable pragmatics of doing science involves the establishment of conceptual "boxes" within which the empirical and theoretical dramas are enacted. If you have no boxes, you are doomed to turn and turn again to each new element that your imagination encompasses. Down this road lies chaos. If your boxes are too tightly circumscribed, you will miss the interactive role that your specimen under examination plays with things not in your box and will learn nothing except the isolated contents. This is as futile as the chaos. Study mass exclusively, and you would never know that there are important relations with energy; study language *in vacuo,* and you will miss the role of language in thought and social interaction.

If you take the brain (or the mind) and box it as a distinct entity, you will, at best, be examining a circumscribed device that exists nowhere. This was Descartes's basic error, the futility and fallaciousness of the mind-body (or, if you prefer, brain-body) dualism. Brains and minds do not occur without bodies and bodies do not occur independent of environments, social, physical, and symbolic. The brain may be viewed as a computational device in some metaphoric sense, and it may be viewed as preprogrammed with innate knowledge in a different metaphoric sense, but it most certainly does not function like either of these in its interactive role with the body of which it is so intimate a part.

This seems so obvious to me that I often have difficulty in comprehending why we should take seriously the strong entailments of the kind of content nativism that people like Chomsky have argued for. I find myself drawn to the conclusion that the rationalist, mentalist program of the content-specific nativist is a homolog of the extreme environmentalism of the behaviorist. They would take us to the same point by different routes. Each would block off psychology, bound the field by affiliation with one particular domain and denial of, at or at best by giving short shrift to, the role of other domains. Both, in their idiosyncratic ways, would actually extend the perceived boundaries of explication beyond what can be conveniently subsumed within their spheres. There is a clear but compelling paradox here; theoretical systems that articulate narrow "boxes" simultaneously invite unwarranted extensions of themselves. By minimizing the roles of learning and the environment (as in the case of content-specific nativism) one, perforce, places excessive emphasis on genetics and biology to the point where one goes beyond the realm where the explanatory power of the model is actually to be found. By minimizing the role of genetics and mentalism (as in the case of radical behaviorism), one yields to the same temptation and commits the parallel sin of attempting to extend the approach well beyond its natural domain.

Functionalism, with its process-specific nativism and its affiliation with interactionism, has built-in safeguards against the kinds of unprofitable paths that either of these two extreme positions typically have taken. Despite the anathematizations of some (see Chomsky, 1977; especially chapter 4), a functionally oriented empiricism is really not so bad. It has always carried with it a number of useful heuristics for dealing with Bronowski's box problem and the pragmatic question of the boundaries of applicability of theory. It is an approach that rarely looks for "the" underlying basis of a particular phenomenon or effect. Rather, it articulates effects as occurring in particular contexts owing to the functions that are fulfilled there.

When Dewey argued back in 1916 against any variety of biological determinism, his prime target was recapitulation theory, but the arguments he presented are perfectly parallel with the ones I have tried to develop here. Determinism, he maintained, should always be resisted when there is a lack of persuasive evidence. Not a lack of subtle argument or a lack of supposedly informed consensus but a lack of incontrovertible evidence. There is much to commend this point of view, and there is much to recommend this position as the prime heuristic for contemporary cognitive psychology.

Finally, let me be as clear as I can about the relationship between the approach we have taken to the study of the cognitive unconscious and the issues discussed in this section. In carrying out the research program into the nature of implicit cognitive processes, my colleagues and I have most assuredly adopted a fairly strong version of Bronowski's "box" strategy. Sharply circumscribed experimental domains were developed such as the artificial grammar-learning technique and the probability-learning experiment. Data bases were built up that had little or nothing to do with any other functional aspect of the behavior of our subjects. And only recently have modest attempts been made to link these implicit processes with other elements of human behavior such as the emotional and neuroanatomical states of our subjects. But these decisions were made pragmatically to ensure that we had a relatively clean domain within which to examine the processes and systems of interest. They were made within the framework of an interactive functionalism, and we have always viewed the processes under examination as merely a small part of the operations of a rich and complex organism. It is my sense that such a research program could not have emerged from a content-specific nativism with its inherent conservatism and its tendency to encapsulate and isolate cognitive processes.

Afterword

In this chapter I tried to cover in some detail a number of issues that were implicated by the material outlined in the earlier chapters. Admittedly, in some cases the presentation was a bit looser than that presented in previous chapters, but since these are, by and large, issues that have less in the way of an empirical literature behind them, I felt that it was legitimate to engage in a little speculation. There are many other ancillary issues that could be introduced here—but I will resist the temptation. Most of them are of a tangential nature, and if I were to begin to try to develop them, the speculation would soon cease to be within acceptable limits. However, I must confess I have a certain fondness for some of them. Let me just mention a few of them as a way of ending this book.

INTUITION. It has always struck me that the basic principles embodied in the standard characterization of implicit learning are very close to what the typical layperson thinks of when the topic of *intuition* comes up. Most people think of intuition as a kind of natural judgment process that takes place without conscious thought and generally outside of any explicit awareness of the knowledge base that allows for that thought. Indeed, I have typically found it easier to talk with intelligent laypersons and scholars from other disciplines about my work than to discuss it with cognitive psychologists. Typically, the nonspecialist finds a common epistemological ground with me as soon as I tell them that my research interest is *intuition*. Perhaps the reason why nonspecialists are less resistant to the entailments of implicit learning is that they have, by and large, not been acculturated with the vestiges of Lockean and Cartesian thought concerning the notions of a cognitive unconscious that encodes abstract structures.

ZEN BUDDHISM. I have long had a lingering sense that the essential features of implicit learning are reflected in the body of thought of Zen Buddhism. Much of the philosophy of Zen is captured by the principles that have emerged from our studies of implicit induction. Specifically, in both, knowledge is assumed to be acquired by an attentive immersion in the subject matter under consideration and, in both, these cognitive processes are viewed as being facilitated by refraining from conscious attempts to fathom the nature of the world about. True understanding is presumed to emerge from long and measured involvement with one's environment, an immersion that allows for appropriate action and sensible functioning within that environment. True understanding, however, is not easily communicated. Indeed, it is maintained within the standard guidelines of Zen teaching that true understanding can only be achieved by the process of immersion; the learning that comes from explicit instruction is regarded as superficial.

PEDAGOGY. There are many elements of the approach that I have taken to implicit learning that suggest a tilt toward the kinds of educational programs championed by John Dewey. Unlike contemporary approaches to pedagogy and instruction, the results from the studies on implicit learning suggest that school curricula should be modified to include more exposure to the variations that the specific subject matter displays and less energy and time should be spent on specific tutoring of rules and formulas. Note that I would not recommend a fully "progressive" educational system; the data do not support such a structure. Recall that several of our experiments (e.g., Reber et al., 1980; Reber & Millward, 1968) indicate that maximal learning takes place when there is some direction provided at the outset about the underlying nature of the environment. However, this explicit element has little or no educational effect without the extended immersion in the stimulus display. We do not learn about the underlying structure of complex environments by explicit instruction; we must experience the patterns of covariation for ourselves.

CULTURAL SHIFTS. In addition to these elements of cognition, I have never been able to shake the feeling that implicit processing is a metaphor for the development and acceptance of new forms of art, music, and fashion within a culture. Typically, when new forms of creative expression are introduced into a given culture, there is stern resistance to them; they are usually subjected to vigorous criticism that often ends up in ridicule. However, with continued exposure to these new forms and with the passage of time, they often find themselves becoming integrated into the culture and recognized as legitimate aesthetic contributions. My gloss on this sociocultural process is that as the perceivers of these new forms are immersed in the variations that the artists, composers, and designers have engendered them with, they gradually induce the underlying structural forms that are there. This process of induction allows the perceivers to develop an abstract representation that reflects the structure with which the creator of these forms has imbued them. Gradually these forms come to be recognized as representing a particular art form; they feel appropriate to the observer who is now in a position to make sensible judgments about them.

Although it has been suggested, notably by Zajonc, that simple familiarity effects can account for this, I do not believe that any characterization that fails to take account of structural components can succeed here. For example, novel forms that are introduced that have little in the way of deep underlying structure to them, such as the new forms of popular music that are introduced every other week, tend not to become fixed in the culture despite the high frequency with which they are displayed. When there is little in the way of underlying structure, there will be little in the way of aesthetic value. They are forms that are too easily learned and hence will have little of lasting value. We did not come to first accept and then admire Jackson Pollack merely because we saw so many examples of his paintings; we had to induce an underlying representation that captured the rich structure displayed in his works.

ANALOGS WITH DOING SCIENCE. This issue is one that I am borrowing from the physician cum physical chemist cum philosopher Michael Polanyi. Polanyi regarded the creative actions of the scientist as essentially unconscious. One of his favorite lines was that those working within a complex environment "knew more than they could say." I first discovered Polanyi during the 1960s when I began working on the general problem of implicit learning. I was struck then at the remarkable parallels between Polanyi's speculative cognitive theories and the results that were coming from these early experiments. Polanyi used the term *tacit knowledge* to refer to the information that was gleaned unconsciously from the stimulus display within which one was immersed. Elsewhere (Reber, in press) I have developed in detail the links between Polanyi's perspective and the one that is developed in this book. Suffice it to say that the parallels between these two approaches suggest that there is much of a real-world nature in the characterization of implicit learning that I have put forward that could be developed by extending the exploration of Polanyi's perspective.

In summary

In this volume I have tried to put forward a global picture of the cognitive unconscious with a focus on the problem of implicit learning. My aims have been (1) to outline the essential features of implicit learning that have emerged from the many studies that have been carried out in a variety of experimental laboratories over the past several decades; (2) to present the various alternative perspectives on this issue that have been proposed by other researchers and to try to accommodate these views with my own; (3) to structure the literature so that it can be seen to dovetail gently within the standard heuristics of evolutionary biology; (4) to present the material within the functionalist approach that has dominated my thinking and, in so doing, to try to show why the experimental data should be seen as entailing particular epistemological perspectives; and (5) to present implicit processing as encompassing a general and ubiquitous set of operations that have wide currency and a goodly number of possible applications. I don't know whether I've managed to pull these off, but I have had fun trying, and I'll take my cue from one of my favorite psychologists, Edward Chace Tolman, whose last scholarly work ended with these wonderful words:

> Since all the sciences, and especially psychology, are still immersed in such tremendous realms of the uncertain and the unknown, the best that any individual scientist, especially any psychologist, can do seems to be to follow his own gleam and his own bent, however inadequate they may be. In the end, the only sure criterion is to have fun (Tolman, 1959, p. 152).

References

Aaronson, D., & Scarborough, H. S. (1977). Performance theories for sentence coding: Some quantitative models. *Journal of Verbal Learning and Verbal Behavior, 16,* 277–303.

Abrams, M. (1987). *Implicit learning in the psychiatrically impaired.* Unpublished doctoral dissertation, City University of New York.

Abrams, M., & Reber, A. S. (1988). Implicit learning: Robustness in the face of psychiatric disorders. *Journal of Psycholinguistic Research, 17,* 425–439.

Allen, R., & Reber, A. S. (1980). Very long term memory for tacit knowledge. *Cognition, 8,* 175–185.

Anderson, J. R. (1976). *Language, memory, and thought.* Hillsdale, NJ: Erlbaum.

——— (1978). Arguments concerning representations for mental imagery. *Psychological Review, 85,* 249–277.

——— (1979). Further arguments concerning representations for mental imagery. *Psychological Review, 86,* 395–406.

——— (1983). *The architecture of cognition.* Cambridge, MA: Harvard University Press.

——— (1987). Skill acquisition: Compilation of weak-method problem solutions. *Psychological Review, 94,* 192–210.

Anderson, J. R., & Bower, G. H. (1973). *Human associative memory.* Washington, DC: Winston.

Anderson, N. H. (1960). Effect of first-order conditional probability in a two-choice learning situation. *Journal of Experimental Psychology, 59,* 73–93.

Andrewsky, E. L., & Seron, X. (1975). Implicit processing of grammatical rules in a classical case of agrammatism. *Cortex, 11,* 379–390.

Baars, B. J. (1986). *The cognitive revolution in psychology.* New York: Guilford.

——— (1988). *A cognitive theory of consciousness.* New York: Cambridge University Press.

Baer, E. K. von (1828). *Entwicklungsgeschichte der thiere: Beobachtung und reflexion.* Königsberg: Bornträger.

Bailey, C. H., & Chen, M. (1991). The anatomy of long-term sensitization in *Aplysia:* Morphological insights into learning and memory. In L. R. Squire, N. M. Weinberger, G. Lynch, & J. L. McGaugh (Eds.), *Memory: Organization and locus of change.* New York: Oxford University Press.

Bauer, R. M. (1984). Autonomic recognition of names and faces in prosopagnosia: A neuropsychological application of the guilty knowledge test. *Neuropsychologia, 22,* 457–469.

Bechtel, W. (1988). *Philosophy of mind: An overview for cognitive science.* Hillsdale, NJ: Erlbaum.

Bergson, H. (1913). *Introduction to metaphysics.* New York: Liberal Arts Press.

Berry, D. C., & Broadbent, D. E. (1984). On the relationship between task performance and associated verbalizable knowledge. *Quarterly Journal of Experimental Psychology, 36,* 209–231.

——— (1987). Explanation and verbalization in a computer-assisted search task. *Quarterly Journal of Experimental Psychology, 39A,* 585–609.

——— (1988). Interactive tasks and the implicit-explicit distinction. *British Journal of Psychology, 79,* 251–272.

——— (1990). The role of instruction and verbalization in improving performance on complex search tasks. *Behavior and Information Technology, 9,* 175–190.

Berry, D. C., & Dienes, Z. (1991). The relationship between implicit memory and implicit learning. *British Journal of Psychology, 82,* 359–373.

Bialystok, E. (1981). Some evidence for the integrity and interaction of two knowledge sources. In R. Anderson (Ed.), *New directions in research on the acquisition and use of a second language.* Rowley, MA: Newbury Press.

Block, N. (1980). Introduction: What is functionalism? In N. Block (Ed.), *Readings in the philosophy of psychology,* Vol. 1 (pp. 171–184). Cambridge, MA: Harvard University Press.

Blumenthal, A. L. (1987). The emergence of psycholinguistics. *Synthese, 72,* 313–324.

Blumstein, S. E., Milberg, W., & Shrier, R. (1982). Semantic processing in aphasia: Evidence from an auditory lexical decision task. *Brain and Language, 17,* 301–315.

Boring, E. G. (1950). *A history of experimental psychology* (2nd Ed.). New York: Appleton, Century, Crofts.

Bowers, K. S., & Meichenbaum, D. (1984) (Eds.). *The unconscious reconsidered.* New York: Wiley.

Bowles, N. L., & Poon, L. W. (1985). Aging and retrieval of words in semantic memory. *Journal of Gerontology, 40,* 71–77.

Bradshaw, J. L. (1974). Peripherally presented and unreported words may bias the perceived meaning of a centrally fixated homograph. *Journal of Experimental Psychology, 103,* 1200–1202.

Braine, M. D. S. (1971). On two types of models of the internalization of grammars. In D. I. Slobin (Ed.), *The ontogenesis of grammar: A theoretical symposium.* New York: Academic Press.

Braine, M. D. S., Brody, R. E., Brooks, P. J., Sudhalter, V., Ross, J. A., Catalano, L., & Fisch, S. M. (1990). Exploring language acquisition in children with a miniature artificial language: Effects of item and pattern frequency, arbitrary subclasses, and correction. *Journal of Memory and Language, 29,* 591–610.

Brainerd, C. J., & Reyna, V. F. (1990). Gist is the grist: Fuzzy-trace theory and the new intuitionism. *Developmental Review, 10,* 3–47.

Brewer, W. F. (1974). There is no convincing evidence for operant or classical conditioning in adult humans. In W. B. Weimer & D. S. Palermo (Eds.), *Cognition and the symbolic processes* (pp. 1–42). Hillsdale, NJ: Erlbaum.

Broadbent, D. E. (1977). Levels, hierarchies, and the locus of control. *Quarterly Journal of Experimental Psychology, 29,* 181–201.

Broadbent, D. E., & Aston, B. (1978). Human control of a simulated economic system. *Ergonomics, 21,* 1035–1043.

Broadbent, D. E., FitzGerald, P., & Broadbent, M. H. P. (1986). Implicit and explicit knowledge in the control of complex systems. *British Journal of Psychology, 77,* 33–50.

Brody, N. (1972). *Personality: Research and theory.* New York: Academic Press.

——— (1989). Unconscious learning of rules: Comment on Reber's analysis of implicit learning. *Journal of Experimental Psychology: General, 118,* 236–238.

Bronowski, J. (1978). *Magic science and civilization.* New York: Columbia University Press.

Brooker, B. H., & Cyr, J. J. (1986). Tables for clinicians to use to convert WAIS-R short forms. *Journal of Clinical Psychology, 42,* 983.

Brooks, L. R. (1978). Nonanalytic concept formation and memory for instances. In E. Rosch & B. B. Lloyd (Eds.), *Cognition and categorization.* New York: Wiley.

Brooks, L. R., & Vokey, J. R. (1991). Abstract analogies and abstracted grammars: Comments on Reber (1989) and Mathews et al. (1989). *Journal of Experimental Psychology: General, 120,* 316–323.

Bruner, J. S., Goodnow, J., & Austin, G. (1956). *A study of thinking.* New York: Wiley.

Burke, C. J., & Estes, W. K. (1957). A component model for stimulus variables in discrimination learning. *Psychometrika, 22,* 133–145.

Burke, D. M., & Yee, P. L. (1984). Semantic priming during sentence processing by young and older adults. *Developmental Psychology, 20,* 903–910.

Burke, S. J. (1974). Language acquisition, language learning, and language teaching. *International Review of Applied Linguistics in Language Teaching, 12,* 53–68.

Bush, R. R., & Mosteller, F. (1951). A mathematical model for simple learning. *Psychological Review, 58,* 313–323.

Butters, N. (1989). *Dissociation of implicit memory in dementia.* Paper presented at the meeting of the Psychonomic Society, Atlanta, GA.

Butters, N., Granholm, E., Salmon, D. P., & Grant, I. (1987). Episodic and semantic memory: A comparison of amnesic and demented patients. *Journal of Clinical and Experimental Neuropsychology, 9,* 479–497.

Butters, N., Salmon, D. P., Heindel, W., & Granholm, E. (1988). Episodic, semantic, and procedural memory: Some comparisons of Alzheimer and Huntington disease patients. In R. D. Terry (Ed.), *Aging and the brain* (pp. 63–87). New York: Raven Press.

Cantor, G. W. (1980). *On symbol string-type and the generalizability of the implicit learning paradigm.* Paper presented at meeting of the Eastern Psychological Association, New York, NY.

Carew, T. J., Hawkins, R. D., & Kandel, E. R. (1983). Differential classical conditioning of a defense withdrawal reflex in *Aplysia californica. Science, 219,* 397–400.

Carmody, D. P., Kundel, H. L., & Toto, L. C. (1984). Comparison scans while reading chest images: Taught, but not practiced. *Investigative Radiology, 19,* 462–466.

Carpenter, W. B. (1874). *Principles of mental physiology.* London: John Churchill.

Carroll, M., Byrne, B., & Kirsner, K. (1985). Autobiographical memory and perceptual learning: A developmental study using picture recognition, naming latency, and perceptual identification. *Memory & Cognition, 13,* 273–279.

Ceci, S. J., & Liker, J. K. (1986a). A day at the races: A study of IQ, expertise, and cognitive psychology. *Journal of Experimental Psychology: General, 115,* 255–266.

——— (1986b). Academic and non-academic intelligence: An experimental separation. In R. J. Sternberg & R. K. Wagner (Eds.), *Practical intelligence: Origins of competence in the everyday world* (pp. 119–142). New York: Cambridge University Press.

Cermak, L. S., Talbot, N., Chandler, K., & Wolbarst, L. R. (1985). The perceptual priming phenomenon in amnesia. *Neuropsychologia, 23,* 615–622.

Cheesman, J., & Merikle, P. M. (1986). Word recognition and consciousness. In D. Besner, T. G. Waller, & G. E. Mackinnon (Eds.), *Reading research: Advances in theory and practice,* Vol. 5 (pp. 311–352). New York: Academic Press.

Chomsky, N. (1977). *Language and responsibility.* New York: Pantheon Books.

——— (1980). *Rules and representations.* New York: Columbia University Press.

——— (1986). *Knowledge of language: Its nature, origin, and use.* New York: Praeger.

Chomsky, N., & Miller, G. A. (1958). Finite state languages. *Information and Control, 1,* 91–112.

Churchland, P. M. (1986). Some reductive strategies in cognitive neurobiology. *Mind, 95,* 279–309.

Churchland, P. S. (1986). *Neurophilosophy: Toward a unified science of the mind-brain.* Cambridge, MA: MIT Press/Bradford Books.

Claparède, E. (1911/1951). Recognition and "me-ness." In D. Rapaport (Ed.), *Organi-*

zation and pathology of thought (pp. 58–75). New York: Columbia University Press. (Reprinted from Reconnaissance et moiité. *Archives de Psychologie, 11,* 79–90.)

Cleeremans, A. (in press-a). The representation of structure in sequence-prediction tasks. *Attention and Performance.*

——— (in press-b). *Mechanisms of implicit learning: Connectionist models of sequence processing.* Cambridge, MA: MIT Press.

Cleeremans, A., & McClelland, J. L. (1991). Learning the structure of event sequences. *Journal of Experimental Psychology: General, 120,* 235–253.

Cleeremans, A., Servan-Schreiber, D., & McClelland, J. L. (1989). Finite state automata and simple recurrent networks. *Neural Computation, 1,* 372–381.

——— (in press). Graded state machines: The representation of temporal contingencies in simple recurrent networks. In Y. Chauvin & D. E. Rumelhart (Eds.), *Back-propogation: Theory, architectures, and applications.* Hillsdale, NJ: Erlbaum.

Cohen, G., & Faulkner, D. (1983). Word recognition: Age differences in contextual facilitation effects. *British Journal of Psychology, 74,* 239–251.

Cohen, N. J. (1984). Preserved learning capacity in amnesia: Evidence for multiple memory systems. In L. R. Squire & N. Butters (Eds.), *Neuropsychology of memory* (pp. 83–102). New York: Guilford.

Cohen, N. J., Ivry, R., & Keele, S. W. (1990). Attention and structure in sequence learning. *Journal of Experimental Psychology: Learning, Memory, and Cognition, 16,* 17–30.

Corkin, S. (1968). Acquisition of motor skill after bilateral medial temporal lobe excision. *Neuropsychologia, 6,* 225–265.

Corteen, R. S., & Wood, B. (1972). Autonomic responses to shock-associated words in an unattended channel. *Journal of Experimental Psychology, 94,* 308–313.

Coslett, H. B. (1986). *Preservation of lexical access in alexia without agraphia.* Paper presented at the 9th European conference of the International Neuropsychological Society, Veldhoven, Netherlands.

Croce, B. (1972). *Aesthetic* (2d Ed.). New York: Norwood Press.

Damasio, A. R., Eslinger, P. J., Damasio, H., Van Hoesen, G. W., & Cornell, S. (1985). Multimodal amnesic syndrome following bilateral temporal and basal forebrain damage. *Archives of Neurology, 42,* 252–259.

Danks, J. H., & Gans, D. L. (1975). Acquisition and utilization of a rule structure. *Journal of Experimental Psychology: Human Learning and Memory, 1,* 201–208.

Darwin, C. (1871). *The descent of man.* London: John Murray.

Davis, B. D. (1986). *Storm over biology: Essays on science, sentiment, and public policy.* Buffalo, NY: Prometheus Books.

De Groot, A. D. (1965). *Thought and choice in chess.* The Hague: Mouton.

De Haan, E. H. F., Young, A. W., & Newcombe, F. (1987). Face recognition without awareness. *Cognitive Neuropsychology, 4,* 385–415.

Dennett, D. C. (1987). Consciousness. In G. L. Gregory (Ed.), *The Oxford companion to the mind* (pp. 161–164). New York: Oxford University Press.

Dennett, D. C. (1991). *Consciousness explained.* Boston: Little, Brown and Co.

Derks, P. L. (1963). Effect of run length in the gambler's fallacy. *Journal of Experimental Psychology, 65,* 213–214.

Dewey, J. (1916). *Democracy and education.* New York: Macmillan.

Diamond, R., & Rozin, P. (1984). Activation of existing memories in the amnesic syndrome. *Journal of Abnormal Psychology, 93,* 98–105.

Dienes, Z. (1992). Connectionist and memory array models of artificial grammar learning. *Cognitive Science, 16,* 41–79.

Dienes, Z., Broadbent, D., & Berry, D. (1991). Implicit and explicit knowledge bases in artificial grammar learning. *Journal of Experimental Psychology: Learning, Memory, and Cognition, 17,* 875–887.

Dixon, N. F. (1971). *Subliminal perception: The nature of a controversy.* New York: McGraw-Hill.

Dixon, N. F. (1981). *Preconscious processing.* New York: Wiley.

Dodwell, P. C. (1975). Contemporary theoretical problems in seeing. In E. C. Carterette & M. P. Friedman (Eds.), *Handbook of perception* Vol. V. New York: Academic Press.

Doppelt, J. E. (1956). Estimating the full score on the Wechsler Adult Intelligence Scale from scores on four subtests. *Journal of Counseling Psychology, 20,* 63–66.

Druhan, B. B., & Mathews, R. C. (1989). THIYOS: A classifier system model of implicit knowledge of artificial grammars. *Proceedings of the 11th annual conference of the Cognitive Science Society.* Hillsdale, NJ: Erlbaum.

Dulany, D. E. (1991). Conscious representation and thought systems. In R. S. Wyer, Jr., & T. K. Srull (Eds.), *Advances in social cognition,* Vol. 4. Hillsdale, NJ: Erlbaum.

Dulany, D. E., Carlson, R. A., & Dewey, G. I. (1984). A case of syntactical learning and judgment: How conscious and how abstract? *Journal of Experimental Psychology: General, 113,* 541–555.

Dulany, D. E., Carlson, R. A., & Dewey, G. I. (1985). On consciousness in syntactical learning and judgment: A reply to Reber, Allen, & Regan. *Journal of Experimental Psychology: General, 114,* 25–32.

Ebbinghaus, H. (1964). *Memory* (Translated from the German edition [1885] by H. A. Ruger and C. E. Bussenius). New York: Dover.

Eimas, P. K., Miller, J. L., & Jusczyk, P. W. (1987). On infant speech perception and the acquisition of language. In S. Harnard (Ed.), *Categorical perception.* New York: Cambridge University Press.

Ellenberger, H. F. (1970). *The discovery of the unconscious.* New York: Basic Books.

Elman, J. L. (1990). Finding structure in time. *Cognitive Science, 14,* 179–211.

Erdelyi, M. H. (1974). A new look at the New Look: Perceptual defense and vigilance. *Psychological Review, 81,* 1–25.

——— (1985). *Psychoanalysis: Freud's cognitive psychology.* New York: Freeman & Co.

——— (1986). Experimental indeterminacies in the dissociation paradigm of subliminal perception: Comment on Holender (1986). *The Behavioral and Brain Sciences, 9,* 30–31.

——— (1992). Psychodynamics and the unconscious. *American Psychologist, 47,* 784–787.

Eriksen, C. W. (1958). Unconscious processes. In M. R. Jones (Ed.), *Nebraska symposium on motivation* (pp. 169–278). Lincoln: University of Nebraska Press.

——— (1960). Discrimination and learning without awareness: A methodological survey and evaluation. *Psychological Review, 67,* 279–300.

Estes, W. K. (1950). Toward a statistical theory of learning. *Psychological Review, 57,* 94–107.

——— (1959). The statistical approach to learning theory. In S. Koch (Ed.), *Psychology: A study of a science,* Vol. 2 (pp. 380–491). New York: McGraw-Hill.

——— (1964). Probability learning. In A. W. Melton (Ed.), *Categories of human learning* (pp. 89–128). New York: Academic Press.

Estes, W. K., & Straughn, J. H. (1954). Analysis of a verbal conditioning situation in terms of statistical learning theory. *Journal of Experimental Psychology, 47,* 225–234.

Feldman, J. (1963). Simulation of behavior in the binary choice experiment. In E. A. Feigenbaum & J. Feldman (Eds.), *Computers and thought* (pp. 329–346). New York: McGraw-Hill.

Flavell, J. H., & Wellman, H. M. (1977). Metamemory. In R. V. Kail & J. W. Hagen

(Eds.), *Perspectives on the development of memory and cognition.* Hillsdale, NJ: Erlbaum.

Fodor, J. A. (1975). *The language of thought.* Cambridge, MA: Harvard University Press.

——— (1983). *Modularity of mind: An essay on faculty psychology.* Cambridge, MA: MIT Press.

Fowler, C., Wolford, G., Slade, R., & Tassinary, L. (1981). Lexical access with and without awareness. *Journal of Experimental Psychology: General, 110,* 341–362.

Fried, L. S., & Holyoak, K. J. (1984). Induction of category distributions: A framework for classification learning. *Journal of Experimental Psychology: Learning, Memory, and Cognition, 10,* 234–257.

Friedman, M. P., Burke, C. J., Cole, M., Keller, L., Millward, R. B., & Estes, W. K. (1964). Two-choice behavior under extended training with shifting probabilities of reinforcement. In R. C. Atkinson (Ed.), *Studies in mathematical psychology.* Stanford, CA: Stanford University Press.

Gabrieli, J. D. E., Keane, M. M., & Corkin, S. (1987). Acquisition of problem-solving skills in global amnesia. *Society for Neurosciences Abstracts, 13,* 1455.

Gardner, H. (1983). *Frames of mind.* New York: Basic Books.

Garner, W. R. (1974). *The processing of information and structure.* Hillsdale, NJ: Erlbaum.

——— (1978). Aspects of a stimulus: Features, dimensions, and configurations. In E. Rosch & B. B. Lloyd (Eds.), *Cognition and categorization.* Hillsdale, NJ: Erlbaum.

Gibbon, J., & Balsam, P. (1981). Spreading association in time. In C. M. Locurto, H. S. Terrace, & J. Gibbon (Eds.), *Autoshaping and conditioning theory* (pp. 219–253). New York: Academic Press.

Gibson, E. J. (1969). *Perceptual learning and development.* New York: Appleton-Century-Crofts.

Gibson, J. J. (1966). *The senses considered as perceptual systems.* Boston: Houghton Mifflin.

——— (1979). *The ecological approach to visual perception.* Boston: Houghton Mifflin.

Glaser, R. (1990). The reemergence of learning theory with instructional research. *American Psychologist, 45,* 29–39.

Gleitman, H. (1981). *Psychology.* New York: J. J. Norton.

Gleitman, L. R., & Wanner, E. (1982). Language acquisition: The state of the art. In E. Wanner & L. R. Gleitman (Eds.), *Language acquisition: The state of the art.* New York: Cambridge University Press.

Glisky, E. L., & Schacter, D. L. (1989). Extending the limits of complex learning in organic amnesia: Computer training in a vocational domain. *Neuropsychologia, 27,* 107–120.

Glisky, E. L., Schacter, D. L., & Tulving, E. (1986). Computer learning by memory-impaired patients: Acquisition and retention of complex knowledge. *Neuropsychologia, 24,* 313–328.

Gluck, M. A., & Bower, G. H. (1988). From conditioning to category learning: An adaptive network model. *Journal of Experimental Psychology: General, 117,* 225–244.

——— (1990). Component and pattern information in adaptive networks. *Journal of Experimental Psychology: General, 119,* 105–109.

Goodnow, J. J. (1955). Response sequences in a pair of two-choice situations. *American Journal of Psychology, 68,* 624–630.

Gordon, P. C., & Holyoak, K. J. (1983). Implicit learning and generalization of the "mere exposure" effect. *Journal of Personality and Social Psychology, 45,* 492–500.

Gould, S. J. (1977a). *Ever since Darwin.* New York: Norton.

——— (1977b). *Ontogeny and phylogeny.* Cambridge, MA: Harvard University Press.

Gould, S. J., & Lewontin, R. C. (1979). The spandrels of San Marco and the Panglossian paradigm. *Royal Society of London: Proceedings B, 205,* 581–598.

Gould, S. J., & Vrba, E. S. (1982). Exaptation–A missing term in the science of form. *Paleobiology, 8,* 4–15.

Graf, P., & Mandler, G. (1984). Activation makes words more accessible but not necessarily more retrievable. *Journal of Verbal Learning and Verbal Behavior, 23,* 553–568.

Graf, P., Mandler, G., & Haden, P. (1982). Simulating amnesic symptoms in normal subjects. *Science, 218,* 1243–1244.

Graf, P., & Schacter, D. L. (1985). Implicit and explicit memory for new associations in normal and amnesic subjects. *Journal of Experimental Psychology: Learning, Memory, and Cognition, 11,* 501–518.

——— (1987). Selective effects of interference on implicit and explicit memory for new associations. *Journal of Experimental Psychology: Learning, Memory, and Cognition, 13,* 45–53.

Graf, P., Squire, L. R., & Mandler, G. (1984). The information that amnesic patients do not forget. *Journal of Experimental Psychology: Learning, Memory, and Cognition, 10,* 164–178.

Gray, J. A. (1982). *The neuropsychology of anxiety: An enquiry into the functions of the septo-hippocampal system.* New York: Oxford University Press.

——— (1984). The hippocampus as an interface between cognition and emotion. In H. L. Roitblat, T. G. Bever, & H. S. Terrace (Eds.), *Animal cognition.* Hillsdale, NJ: Erlbaum.

Green, R. E. A., & Shanks, D. R. (in press). On the existence of independent learning systems: An examination of some evidence. *Memory and Cognition.*

Greenbaum, J. L., & Graf, P. (1989). Preschool period development of implicit and explicit remembering. *Bulletin of the Psychonomic Society, 27,* 417–420.

Greenspoon, J. (1955). The reinforcing effects of two spoken sounds on the frequency of two responses. *American Journal of Psychology, 68,* 409–416.

Greenwald, A. G. (1992). New look 3: Unconscious cognition reclaimed. *American Psychologist, 47,* 766–779.

Gregory, R. L. (1987) (Ed.). *The Oxford companion to the mind.* New York: Oxford University Press.

Griffin, D. R. (1981). *The question of animal awareness: Evolutionary continuity of mental experience.* New York: Rockefeller University Press.

——— (1984). *Animal thinking.* Cambridge, MA: Harvard University Press.

Haith, M. M., Hazan, C., & Goodman, G. S. (1988). Expectation and anticipation of dynamic visual events by 3.5-month-old babies. *Child Development, 59,* 467–479.

Haith, M. M., & McCarty, M. E. (1990). Stability of visual expectations at 3.0 months of age. *Developmental Psychology, 26,* 68–74.

Hasher, L., & Chromiak, W. (1977). The processing of frequency information: An automatic mechanism? *Journal of Verbal Learning and Verbal Behavior, 16,* 173–184.

Hasher, L., & Zacks, R. T. (1979). Automatic and effortful processes in memory. *Journal of Experimental Psychology: General, 108,* 356–388.

——— (1984). Automatic processing of fundamental information. *American Psychologist, 39,* 1372–1388.

Hayek, F. A. von (1962). Rules, perception, and intelligibility. *Proceedings of the British Academy, 48,* 321–344.

Hayes, N. A., & Broadbent, D. E. (1988). Two modes of learning for interactive tasks. *Cognition, 28,* 249–276.

Hayes-Roth, F. (1979). Distinguishing theories of representation: A critique of Anderson's "Arguments concerning mental imagery." *Psychological Review, 86,* 376–392.

Heindel, W. C., Butters, N., & Salmon, D. P. (1988). Impaired learning of a motor skill in patients with Huntington's disease. *Behavioral Neuroscience, 102,* 141–147.

Helmholtz, H. (1867/1962). *Treatise on Physiological Optics, Vol. 3* (Translated from the 3d German edition [1867], J. P. C. Southall, Ed.). New York: Dover.

Hering, E. (1920). Memory as a universal function of organized matter. In S. Butler (Ed.), *Unconscious memory* (pp. 63–86). London: Jonathan Cape.

Heron, A., & Chown, S. (1967). *Age and function.* Boston: Little, Brown, & Co.

Hill, T., Lewicki, P., Czyzewska, M., & Boss, A. (1989). Self-perpetuating development of encoding biases in person perception. *Journal of Personality and Social Psychology, 57,* 373–387.

Hirsh, R. (1980). The hippocampus, conditional operations, and cognition. *Physiological Psychology, 8,* 175–182.

Hirshman, E., Snodgrass, J. G., Mindes, J., & Feenan, K. (1990). Conceptual priming in picture fragment completion. *Journal of Experimental Psychology: Learning, Memory, and Cognition, 16,* 634–647.

Hogarth, R. (1987). *Judgment and choice* (2d ed.). New York: Wiley.

Holender, D. (1986). Semantic activation without conscious identification in dichotic listening, parafoveal vision, and visual masking: A survey and appraisal. *The Behavioral and Brain Sciences, 9,* 1–66.

Holland, J. H., Holyoak, K. J., Nisbett, R. E., & Thagard, P. R. (1986). *Induction: Processes of inference, learning and discovery.* Cambridge, MA: MIT Press.

Honig, W. K. (1978). Studies of working memory in the pigeon. In S. H. Hulse, H. Fowler, & W. K. Honig (Eds.), *Cognitive processes in animal behavior* (pp. 211–248). Hillsdale, NJ: Erlbaum.

Howard, D. V., Lasaga, M. I., & McAndrews, M. P. (1980). Semantic activation during memory encoding across the adult life span. *Journal of Gerontology, 35,* 884–890.

Howard, J. H., & Ballas, J. A. (1980). Syntactic and semantic factors in the classification of nonspeech transient patterns. *Perception and Psychophysics, 28,* 431–439.

——— (1982). Acquisition of acoustic pattern categories by exemplar observation. *Organizational Behavior and Human Performance, 30,* 157–182.

Howard, J. H., Mutter, S. A., & Howard, D. V. (1992). Serial pattern learning by event observation. *Journal of Experimental Psychology: Learning, Memory, & Cognition, 18,* 1029–1040.

Hull, C. L. (1920). Quantitative aspects of the evolution of concepts. *Psychological Monographs,* Whole No. 123.

Humphreys, L. G. (1939). Acquisition and extinction of verbal expectations in a situation analogous to conditioning. *Journal of Experimental Psychology, 25,* 294–301.

Jacobs, W. J., & Nadel, L. (1985). Stress induced recovery of fears and phobias. *Psychological Review, 92,* 512–531.

Jacoby, L. L. (1983). Perceptual enhancement: Persistent effects of an experience. *Journal of Experimental Psychology: Learning, Memory, and Cognition, 9,* 21–38.

——— (1984). Incidental versus intentional retrieval: Remembering and awareness as separate issues. In L. R. Squire & N. Butters (Eds.), *Neuropsychology of memory* (pp. 145–156). New York: Guilford Press.

——— (1992). *Strategic versus automatic influences of memory: Attention, awareness, and control.* Paper presented at the meeting of the Psychonomic Society, St. Louis, Missouri.

Jacoby, L. L., & Dallas, M. (1981). On the relationship between autobiographical memory and perceptual learning. *Journal of Experimental Psychology: General, 110,* 306–340.

Jacoby, L. L., Lindsay, D. S., & Toth, J. P. (1992). Unconscious influences revealed. *American Psychologist, 47,* 802–809.

Jacoby, L. L., & Whitehouse, K. (1989). An illusion of memory: False recognition influenced by unconscious perception. *Journal of Experimental Psychology: General, 118,* 126–135.

Jacoby, L. L., & Witherspoon, D. (1982). Remembering without awareness. *Canadian Journal of Psychology, 36,* 300–324.

Jakobson, R. (1968). *Child language aphasia and phonological universals.* The Hague: Mouton & Co.

James, W. (1890). *The principles of psychology.* New York: Holt.

Jarvik, M. E. (1951). Probability learning and a negative recency effect in the serial anticipation of alternative symbols. *Journal of Experimental Psychology, 41,* 291–297.

Jaynes, J. (1976). *The origins of consciousness in the breakdown of the bicameral mind.* Boston: Houghton Mifflin.

Jenkins, H. M. (1984). Time and contingency in classical conditioning. In J. Gibbon & L. Allen (Eds.), *Timing and time perception, Annals of the New York Academy of Sciences,* Vol. 423 (pp. 242–253). New York: New York Academy of Sciences.

Jenkins, H. M., Barnes, R. A., & Barrera, F. J. (1981). Why autoshaping depends on trial spacing. In C. M. Locurto, H. S. Terrace, & J. Gibbon (Eds.), *Autoshaping and conditioning theory.* New York: Academic Press.

Jenkins, J. G. (1933). Instruction as a factor in "incidental" learning. *American Journal of Psychology, 45,* 471–477.

Jennings, J., & Jacoby, L. L. (1992). *Automatic versus intentional uses of memory: Aging, attention, and control.* Manuscript submitted for publication.

Jennings, P. J., & Keele, S. W. (1991). A computational model of attentional requirements in sequence learning. *Proceedings of the 13th annual conference of the Cognitive Science Society.* Hillsdale, NJ: Erlbaum.

Johnson, M. K., Kim, J. K., & Risse, G. (1985). Do alcoholic Korsakoff's syndrome patients acquire affective reactions? *Journal of Experimental Psychology: Learning, Memory, and Cognition, 11,* 27–36.

Jones, M. R., & Myers, J. L. (1966). A comparison of two methods of event randomization in probability learning. *Journal of Experimental Psychology, 72,* 909–911.

Jordan, M. I. (1986). Attractor dynamics and parallelism in a connectionist sequential machine. *Proceedings of the 8th annual conference of the Cognitive Science Society.* Hillsdale, NJ: Erlbaum.

Jung, C. (1926). *Psychological types* (H. G. Baynes, Trans.). London: Routledge and Kegan Paul.

Kahneman, D., Slovic, P., & Tversky, A. (1982) (Eds.). *Judgment under uncertainty: Heuristics and biases.* New York: Cambridge University Press.

Kahneman, D., & Treisman, A. (1984). Changing views of attention and automaticity. In R. Parasuraman & D. R. Davies (Eds.), *Varieties of attention.* New York: Academic Press.

Kassin, S. M., & Reber, A. S. (1979). Locus of control and the learning of an artificial language. *Journal of Research in Personality, 13,* 111–118.

Kihlstrom, J. F. (1980). Posthypnotic amnesia for recently learned material: Interactions with "episodic" and "semantic" memory. *Cognitive Psychology, 12,* 227–251.

——— (1985). Hypnosis. *Annual Review of Psychology, 36,* 385–418.

——— (1987). The cognitive unconscious. *Science, 237,* 1445–1452.

——— (1990). The psychological unconscious. In H. Pervin (Ed.), *The handbook of personality* (pp. 445–464). New York: Guilford Press.

Knopman, D. S., & Nissen, M. J. (1987). Implicit learning in patients with probable Alzheimer's disease. *Neurology, 37,* 784–488.

Knowlton, B. J., Ramus, S. J., & Squire, L. R. (1992). Intact artificial grammar learning in amnesia: Dissociation of abstract knowledge and memory for specific instances. *Psychological Science, 3,* 172–179.

Korsakoff, S. S. Etude médico-psychologique sur une forme des maladies de la mémoire [Medical-psychological study of a form of memory disorder]. *Revue Philosophique, 28,* 501–530.

Kruschke, J. K. (1992). ALCOVE: An exemplar-based connectionist model of category learning. *Psychological Review, 99,* 22–44.

Kunst-Wilson, W. R., & Zajonc, R. B. (1980). Affective discrimination of stimuli that cannot be recognized. *Science, 207,* 557–558.

Kushner, M. (1992). *Implicit detection of event interdependencies.* Unpublished doctoral dissertation, City University of New York.

Kushner, M., Cleeremans, A., & Reber, A. S. (1991). Implicit detection of event interdependencies and a PDP model of the process. *Proceedings of the 13th annual conference of the Cognitive Science Society.* Hillsdale, NJ: Erlbaum.

Langer, E. (1978). Rethinking the role of thought in social interaction. In J. Harvey, W. Ickes, & R. Kidd (Eds.), *New directions in attribution theory,* Vol. 2 (pp. 35–58). Hillsdale, NJ: Erlbaum.

Langer, E., Blank, A., & Chanowitz, B. (1978). The mindlessness of ostensibly thoughtful action: The role of "placebic" information in interpersonal interaction. *Journal of Personality and Social Psychology, 36,* 635–642.

Lenneberg, E. H. (1967). *Biological foundations of language.* New York: Wiley.

Lewandowsky, S., Dunn, J. C., & Kirsner, K. (1989) (Eds.). *Implicit memory.* Hillsdale, NJ: Erlbaum.

Lewicki, P. (1985). Nonconscious biasing effects of single instances on subsequent judgments. *Journal of Personality and Social Psychology, 48,* 563–574.

——— (1986a). *Nonconscious social information processing.* New York: Academic Press.

——— (1986b). Processing information about covariations that cannot be articulated. *Journal of Experimental Psychology: Learning, Memory, and Cognition, 12,* 135–146.

Lewicki, P., Czyzewska, M., & Hoffman, H. (1987). Unconscious acquisition of complex procedural knowledge. *Journal of Experimental Psychology: Learning, Memory, and Cognition, 13,* 523–530.

Lewicki, P., & Hill, T. (1987). Unconscious processes as explanations of behavior in cognitive, personality, and social psychology. *Personality and Social Psychology Bulletin, 13,* 355–362.

——— (1989). On the status of nonconscious processes in human cognition: Comment on Reber. *Journal of Experimental Psychology: General, 118,* 239–241.

Lewicki, P., Hill, T., & Bizot, E. (1988). Acquisition of procedural knowledge about a pattern of stimuli that cannot be articulated. *Cognitive Psychology, 20,* 24–37.

Lewicki, P., Hill, T., & Sasaki, I. (1989). Self-perpetuating development of encoding biases. *Journal of Experimental Psychology: General, 118,* 323–337.

Lewontin, R. C., Rose, S., & Kamin, L. J. (1984). *Not in our genes: Biology, ideology, and human nature.* New York: Pantheon Books.

Light, L. L., & Singh, A. (1987). Implicit and explicit memory in young and older adults. *Journal of Experimental Psychology: Learning, Memory, and Cognition, 13,* 531–541.

Lightfoot, D. (1982). *The language lottery: Toward a biology of grammars.* Cambridge, MA: MIT Press.

Lindsay, P. H., & Norman, D. A. (1977). *Human Information Processing* (2d Ed.). New York: Academic Press.

Mace, W. M. (1974). Gibson's strategy for perceiving: Ask not what's inside your head but what your head's inside of. In R. Shaw & J. Bransford (Eds.), *Perceiving, acting and knowing*. Hillsdale, NJ: Erlbaum.

Macmillan, N. A. (1986). The psychophysics of subliminal perception: Comment on Holender (1986). *The Behavioral and Brain Sciences, 9,* 38–39.

Manza, L. (1992). *Transfer appropriate processing and the representation of implicitly acquired information.* Paper presented at the annual meeting of the Eastern Psychological Association, Boston, MA.

Manza, L., & Reber, A. S. (1992). *Inter- and intra-modal transfer of an implicitly acquired rule system.* Unpublished manuscript.

Marcel, A. J. (1983a). Conscious and unconscious perception: Experiments on visual masking and word recognition. *Cognitive Psychology, 15,* 197–237.

——— (1983b). Conscious and unconscious perception: An approach to the relations between phenomenal experience and perceptual processes. *Cognitive Psychology, 15,* 238–300.

Mathews, R. C. (1990). Abstractness of implicit grammar knowledge: Comments on Perruchet and Pacteau's analysis of synthetic grammar learning. *Journal of Experimental Psychology: General, 119,* 412–416.

——— (1991). The forgetting algorithm: How fragmentary knowledge of exemplars can yield abstract knowledge. *Journal of Experimental Psychology: General, 120,* 117–119.

Mathews, R. C., Buss, R. R., Chinn, R., & Stanley, W. B. (1988). The role of explicit and implicit learning processes in concept discovery. *Quarterly Journal of Experimental Psychology, 40A,* 135–165.

Mathews, R. C., Buss, R. R., Stanley, W. B., Blanchard-Fields, F., Cho, J.-R., & Druhan, B. (1989). The role of implicit and explicit processes in learning from examples: A synergistic effect. *Journal of Experimental Psychology: Learning, Memory, and Cognition, 15,* 1083–1100.

Mathews, R. C., Druhan, B. B., & Roussel, L. G. (1989). *Forgetting is learning: Evaluation of three induction algorithms for learning artificial grammar.* Paper presented at the annual meeting of the Psychonomic Society, Boston.

McAndrews, M. P., & Moscovitch, M. (1985). Rule-based and exemplar-based classification in artificial grammar learning. *Memory & Cognition, 13,* 469–475.

McClelland, J. L., & Rumelhart, D. E. (1986) (Eds.). *Parallel distributed processing: Explorations in the microstructure of cognition,* Vol. 2. Cambridge, MA: MIT Press.

Medin, D. L. (1989). Concepts and conceptual structure. *American Psychologist, 44,* 1469–1481.

Merikle, P. M., & Cheesman, J. (1986). Consciousness is a "subjective" state. *Behavioral and Brain Sciences, 9,* 42.

Milberg, W., & Blumstein, S. E. (1981). Lexical decision and aphasia: Evidence for semantic processing. *Brain and Language, 14,* 371–385.

Milberg, W., Blumstein, S. E., & Dworetzky, B. (1987). Processing of lexical ambiguities in aphasia. *Brain and Language, 31,* 138–150.

Miller, G. A. (1962). *Psychology: The science of mental life.* New York: Harper & Row.

——— (1967). *The psychology of communication.* New York: Basic Books.

Miller, G. A., & Stein, M. (1963). *Grammarama: I.* Preliminary studies and analysis of protocols. Science Report CS-2, Center for Cognitive Studies, Harvard University, Cambridge, MA.

Miller, R. R., Kasprow, W. J., & Schachtman, T. R. (1986). Retrieval variability: Sources and consequences. *American Journal of Psychology, 99,* 145–218.

Miller, R. R., & Matzel, L. D. (1988). The comparator hypothesis: A response rule for the expression of associations. In G. H. Bower (Ed.), *The psychology of learning and motivation,* Vol. 22 (pp. 51–92). New York: Academic Press.

Millward, R. B. (1981). Models of concept formation. In R. E. Snow, P. A. Frederico, & W. E. Montague (Eds.), *Aptitude, learning, and instruction: Cognitive process analysis.* Hillsdale, NJ: Erlbaum.

Millward, R. B., & Reber, A. S. (1968). Event-recall in probability learning. *Journal of Verbal Learning and Verbal Behavior, 7,* 980–989.

——— (1972). Probability learning: Contingent-event sequences with lags. *American Journal of Psychology, 85,* 81–98.

Milner, B. (1962). Les troubles de la memoire accompagnant des lesion hippocampiques bilaterales [Disorders of memory accompanying bilateral hippocampal lesions]. In *Physiologie de l'hippocampe.* Paris: Centre National de las Recherche Scientifique.

Milner, B., Corkin, S., & Teuber, H. L. (1968). Further analysis of the hippocampal amnesic syndrome: 14 year follow-up study of H. M. *Neuropsychologia, 6,* 215–234.

Mishkin, M., Malamut, B., & Bachevalier, J. (1984). Memories and habits: Two neural systems. In J. L. McGaugh, G. Lynch, & N. M. Weinberger (Eds.), *Neurobiology of learning and memory* (pp. 65–77). New York: Guilford Press.

Moeser, S. D., & Bregman, A. S. (1972). The role of reference in the acquisition of a miniature artificial language. *Journal of Verbal Learning and Verbal Behavior, 11,* 759–769.

Moeser, S. D., & Olson, A. J. (1974). The role of reference in children's acquisition of a miniature artificial language. *Journal of Verbal Learning and Verbal Behavior, 13,* 204–218.

Morgan, C. Lloyd (1894). *An introduction to comparative psychology.* London: Walter Scott, Ltd.

Morgan, J. L., & Newport, E. L. (1981). The role of constituent structure in the induction of an artificial language. *Journal of Verbal Learning and Verbal Behavior, 20,* 67–85.

Morgan, M. J. (1977). *Molyneux's question: Vision, touch and the philosophy of perception.* New York: Cambridge University Press.

Moss, M. B., Albert, M. S., Butters, N., & Payne, M. (1986). Differential papers of memory loss among patients with Alzheimer's disease, Huntington's disease, and alcoholic Korsakoff's syndrome. *Archives of Neurology, 43,* 239–246.

Neimark, E. D., & Estes, W. K. (1967) (Eds.). *Stimulus sampling theory.* San Francisco: Holden-Day.

Neisser, U. (1976). *Cognition and reality.* San Francisco: Freeman & Co.

Nelson, K. (1986). *Making sense: The acquisition of shared meaning.* New York: Academic Press.

Newcombe, F., Young, A. W., & De Haan, E. H. F. (1989). Prosopagnosia and object agnosia without covert recognition. *Neuropsychologia, 27,* 179–191.

Newell, A., & Simon, H. A. (1972). *Human problem solving.* Englewood Cliffs, NJ: Prentice-Hall.

Nisbett, R. E., & Ross, L. (1980). *Human inference: Strategies and shortcomings of social judgment.* Englewood Cliffs, NJ: Prentice-Hall.

Nisbett, R. E., & Wilson, T. D. (1977). Telling more than we know: Verbal reports on mental processes. *Psychological Review, 84,* 231–259.

Nissen, M. J., & Bullemer, P. (1987). Attentional requirements of learning: Evidence from performance measures. *Cognitive Psychology, 19,* 1–32.

Nissen, M. J., Knopman, D. S., & Schacter, D. L. (1987). Neurochemical dissociation of memory systems. *Neurology, 37,* 789–794.

Nissen, M. J., Willingham, D., & Hartman, M. (1989). Explicit and implicit remembering: When is learning preserved in amnesia? *Neuropsychologia, 27,* 341–352.

Osgood, C. E. (1953). *Method and theory in experimental psychology.* New York: Oxford University Press.

Palmer, S. E. (1978). Fundamental aspects of cognitive representation. In E. Rosch & B. B. Lloyd (Eds.), *Cognition and categorization.* Hillsdale, NJ: Erlbaum.

Parkin, A. J., & Streete, S. (1988). Implicit and explicit memory in young and older adults. *British Journal of Psychology, 79,* 361–369.

Perruchet, P., Gallego, J., Savy, I. (1990). A critical reappraisal of the evidence for unconscious abstraction of deterministic rules in complex experimental situation. *Cognitive Psychology, 22,* 493–516.

Perruchet, P., & Pacteau, C. (1990). Synthetic grammar learning: Implicit rule abstraction or explicit fragmentary knowledge? *Journal of Experimental Psychology: General, 119,* 264–275.

Perruchet, P., & Pacteau, C. (1991). Implicit acquisition of abstract knowledge about artificial grammars: Some methodological and conceptual issues. *Journal of Experimental Psychology: General, 120,* 112–116.

Perry, C., & Laurence, J. R. (1984). Mental processing outside of awareness: The contributions of Freud and Janet. In K. S. Bowers & D. Meichenbaum (Eds.), *The unconscious reconsidered* (pp. 9–48). New York: Wiley.

Pfungst, O. (1965). *Clever Hans: The horse of Mr. von Osten* (Edited by R. Rosenthal, originally published, 1911). New York: Holt, Rinehart and Winston.

Piatelli-Palmarini, M. (1980) (Ed.). *Language and learning: The debate between Jean Piaget and Noam Chomsky.* Cambridge, MA: Harvard University Press.

Pinker, S. (1989). Language acquisition. In M. I. Posner (Ed.), *Foundations of cognitive science,* Cambridge, MA: MIT Press.

Polanyi, M. (1958). *Personal knowledge: Toward a post-critical philosophy.* Chicago: University of Chicago Press.

——— (1962). *Personal knowledge: Toward a post-critical philosophy.* Chicago: University of Chicago Press.

——— (1966). *The tacit dimension.* Garden City, NY: Doubleday.

Posner, M. I., & Keele, S. W. (1968). On the genesis of abstract ideas. *Journal of Experimental Psychology, 77,* 353–363.

——— (1970). Retention of abstract ideas. *Journal of Experimental Psychology, 83,* 304–308.

Pylyshyn, Z. (1979). Validating computational models: A critique of Anderson's indeterminacy of representation claim. *Psychological Review, 86,* 383–394.

——— (1980). Computation and cognition: Issues in the foundation of cognitive science. *Behavioral and Brain Sciences, 3,* 111–169.

Rathus, J., Reber, A. S., & Kushner, M. (1990). *Implicit and explicit learning: Differential effects of affective states.* Unpublished manuscript.

Rawlins, J. N. P. (1985). Associations across time: The hippocampus as a temporary memory store. *Behavioral and Brain Sciences, 8,* 479–528.

Razran, G. (1961). The observable unconscious and the inferable conscious in current Soviet psychophysiology. *Psychological Review, 68,* 81–147.

Reber, A. S. (1965). *Implicit learning of artificial grammars.* Unpublished MA thesis, Brown University.

——— (1967a). Implicit learning of artificial grammars. *Journal of Verbal Learning and Verbal Behavior, 6,* 317–327.

——— (1967b). *A perceptual learning analysis of probability learning.* Unpublished doctoral dissertation, Brown University.

——— (1969). Transfer of syntactic structure in synthetic languages. *Journal of Experimental Psychology, 81,* 115–119.

——— (1973). On psycho-linguistic paradigms. *Journal of Psycholinguistic Research, 2,* 289–319.

——— (1976). Implicit learning of synthetic languages: The role of instructional set. *Journal of Experimental Psychology: Human Learning and Memory, 2,* 88–94.

——— (1985). *Dictionary of Psychology.* London: Penguin Books, Ltd.

——— (1987). The rise and (surprisingly rapid) fall of psycholinguistics. *Synthese, 72,* 325–339.

——— (1989a). Implicit learning and tacit knowledge. *Journal of Experimental Psychology: General, 118,* 219–235.

——— (1989b). More thoughts on the unconscious: Reply to Brody and to Lewicki and Hill. *Journal of Experimental Psychology: General, 118,* 242–244.

——— (1990). The primacy of the implicit: A comment on Perruchet and Pacteau. *Journal of Experimental Psychology: General, 119,* 340–342.

——— (1991). *Personal knowledge and the cognitive unconscious.* Address presented at the Centennial Celebration of the birth of Michael Polanyi, Kent, OH.

——— (1992a). An evolutionary context for the cognitive unconscious. *Philosophical Psychology, 5,* 33–51.

——— (1992b). The cognitive unconscious: An evolutionary perspective. *Consciousness and Cognition, 1,* 93–133.

——— (in press). Personal knowledge and the cognitive unconscious. In R. Wilken (Ed.), *Michael Polanyi: A century of thought.*

Reber, A. S., & Allen, R. (1978). Analogy and abstraction strategies in synthetic grammar learning: A functionalist interpretation. *Cognition, 6,* 189–221.

Reber, A. S., Allen, R., & Regan, S. (1985). Syntactical learning and judgment, still unconscious and still abstract: Comment on Dulany, Carlson, & Dewey. *Journal of Experimental Psychology: General, 114,* 17–24.

Reber, A. S., Kassin, S. M., Lewis, S., & Cantor, G. W. (1980). On the relationship between implicit and explicit modes in the learning of a complex rule structure. *Journal of Experimental Psychology: Human Learning and Memory, 6,* 492–502.

Reber, A. S., & Lewis, S. (1977). Toward a theory of implicit learning: The analysis of the form and structure of a body of tacit knowledge. *Cognition, 5,* 333–361.

Reber, A. S., & Millward, R. B. (1965). Probability learning and memory for event sequences. *Psychonomic Science, 3,* 431–432.

——— (1968). Event observation in probability learning. *Journal of Experimental Psychology, 77,* 317–327.

——— (1971). Event tracking in probability learning. *American Journal of Psychology, 84,* 85–99.

Reber, A. S., Walkenfeld, F. F., & Hernstadt, R. (1991). Implicit and explicit learning: Individual differences and IQ. *Journal of Experimental Psychology: Learning, Memory, and Cognition, 17,* 888–896.

Reingold, E. M., & Merikle, P. M. (1988). Using direct and indirect measures to study perception without awareness. *Perception & Psychophysics, 44,* 563–575.

Renault, B., Signoret, J.-L., Debruille, B., Breton, F., & Bolgert, F. (1989). Brain potentials reveal covert facial recognition in prosopagnosia. *Neuropsychologia, 27,* 905–912.

Rescorla, R. A. (1967). Pavlovian conditioning and its proper control procedures. *Psychological Review, 74,* 71–80.

——— (1988). Pavlovian conditioning: It's not what you think it is. *American Psychologist, 43,* 151–160.

Rescorla, R. A., & Wagner, A. R. (1972). A theory of Pavlovian conditioning: Variations in the effectiveness of reinforcement and nonreinforcement. In A. H. Black & W. F.

Prokasy (Eds.), *Classical conditioning II: Current research and theory* (pp. 64–99). New York: Appleton-Century-Crofts.

Ribot, T. (1885). *Diseases of memory.* New York: Appleton.

Richards, R. J. (1987). *Darwin and the emergence of evolutionary theories of mind and behavior.* Chicago: University of Chicago Press.

Roediger, H. L. (1990). Implicit memory: Retention without remembering. *American Psychologist, 45,* 1043–1056.

Romanes, G. J. (1882). *Animal intelligence.* London: Kegan, Paul, Trench & Co.

——— (1883). *Mental evolution in animals.* London: Kegan, Paul, Trench, & Co.

Rosch, E., & Lloyd, B. B. (1978). Representations. In E. Rosch & B. B. Lloyd (Eds.), *Cognition and categorization.* Hillsdale, NJ: Erlbaum.

Roter, A. (1985). *Implicit processing: A developmental study.* Unpublished doctoral dissertation, City University of New York.

Roussel, L. G., Mathews, R. C., & Druhan, B. B. (1990). Rule induction and interference in the absence of feedback: As classifier system model. *Proceedings of the 12th annual conference of the cognitive science society.* Hillsdale, NJ: Erlbaum.

Rovee-Collier, C. (1990). The "memory system" of prelinguistic infants. *Annals of the New York Academy of Sciences, 608,* 517–542.

Roy, F. A. (1982). Action and performance. In A. W. Ellis (Ed.), *Normality and pathology in cognitive functions.* New York: Academic Press.

Rozin, P. (1976). The evolution of intelligence and access to the cognitive unconscious. *Progress in Psychobiology and Physiological Psychology, 6,* 245–280.

Rumelhart, D. E., & McClelland, J. L. (1986) (Eds.). *Parallel distributed processing: Explorations in the microstructure of cognition,* Vol. 1. Cambridge, MA: MIT Press.

Ryle, G. (1949). *The concept of mind.* London: Hutchinson.

Salthouse, T. A. (1982). *Adult cognition.* New York: Springer-Verlag.

Salthouse, T. A., & Somberg, B. L. (1982). Skilled performance: Effects of adult age and experience on elementary processes. *Journal of Experimental Psychology: General, 111,* 176–207.

Sanderson, P. M. (1989). Verbalizable knowledge and skilled task performance: Association, dissociation, and mental models. *Journal of Experimental Psychology: Learning, Memory, and Cognition, 15,* 729–747.

——— (1990). *Implicit and explicit control of a dynamic task: Empirical and conceptual issues.* Technical Report No. EPRL-90-02, University Illinois at Urbana-Champaign, Engineering Psychology Research Laboratory.

Sarason, I. G. (1978). The Test Anxiety Scale: Concept and research. In C. D. Spielberger & I. G. Sarason (Eds.), *Stress and anxiety,* Vol. 2. Washington, DC: Hemisphere.

Scarborough, D. L., Cortese, C., & Scarborough, H. S. (1977). Frequency and repetition effects in lexical memory. *Journal of Experimental Psychology: Human Perception and Performance, 3,* 1–17.

Scarborough, D. L., Gerard, L., & Cortese, C. (1979). Accessing lexical memory: The transfer of word repetition effects across task and modality. *Memory & Cognition, 7,* 3–12.

Schacter, D. L. (1982). *Stranger behind the engram: Theories of memory and the psychology of science.* Hillsdale, NJ: Erlbaum.

——— (1984). Toward the multidisciplinary study of memory: Ontogeny, phylogeny and pathology of memory systems. In L. R. Squire & N. Butters (Eds.), *Neuropsychology of memory* (pp. 13–24). New York: Guilford Press.

——— (1987). Implicit memory: History and current status. *Journal of Experimental Psychology: Learning, Memory, and Cognition, 13,* 501–518.

Schacter, D. L., & Graf, P. (1986). Preserved memory in amnesic patients: Perspectives from research on direct priming. *Journal of Clinical and Experimental Neuropsychology, 8,* 727–743.

——— (1989). Modality specificity of implicit memory for new associations. *Journal of Experimental Psychology: Learning, Memory, and Cognition, 15,* 3–12.

Schacter, D. L., McAndrews, M. P., & Moscovitch, M. (1988). Access to consciousness: Dissociations between implicit and explicit knowledge in neuropsychological syndromes. In L. Weiskrantz (Ed.), *Thought without language.* London: Oxford University Press.

Schacter, D. L., & Moscovitch, M. (1984). Infants, amnesics, and dissociable memory systems. In M. Moscovitch (Ed.), *Infant memory* (pp. 173–216). New York: Plenum.

Schacter, D. L., & Tulving, E. (1982). Memory, amnesia, and the episodic/semantic distinction. In R. L. Isaacson & N. E. Spear (Eds.), *The expression of knowledge* (pp. 33–65). New York: Plenum.

Schank, J. C., & Wimsatt, W. C. (1987). Generative entrenchment and evolution. In A. Fine & P. Machamer (Eds.), *PSA 1986: Proceedings of the meetings of the Philosophy of Science Association,* Vol. 7 (pp. 33–60). East Lansing, MI: Philosophy of Science Association.

Schlesinger, I. M. (1982). *Steps to language: Toward a theory of native language acquisition.* Hillsdale, NJ: Erlbaum.

Seamon, J. G., Brody, N., & Kauff, D. M. (1983). Affective discrimination of stimuli that are not recognized: Effects of shadowing, masking, and cerebral laterality. *Journal of Experimental Psychology: Learning, Memory, and Cognition, 9,* 544–555.

Seamon, J. G., Marsh, R. L., & Brody, N. (1984). Critical importance of exposure duration for affective discrimination of stimuli that are not recognized. *Journal of Experimental Psychology: Learning, Memory, and Cognition, 10,* 465–469.

Servan-Schreiber, E., & Anderson, J. R. (1990). Learning artificial grammars with competitive chunking. *Journal of Experimental Psychology: Learning, Memory, and Cognition, 16,* 592–608.

Shallice, T., & Saffran, E. (1986). Lexical processing in the absence of explicit word identification: Evidence from a letter-by-letter reader. *Cognitive Neuropsychology, 3,* 429–458.

Shanks, D. R. (1985). Continuous monitoring of human contingency judgment across trials. *Memory & Cognition, 13,* 158–167.

——— (1990). Connectionism and human learning: A critique of Gluck and Bower (1988). *Journal of Experimental Psychology: General, 119,* 101–104.

Shaw, R., & Pittinger, J. (1977). Perceiving the face of change in changing faces: Implications for a theory of object perception. In R. Shaw & J. Bransford (Eds.), *Perceiving, acting and knowing.* Hillsdale, NJ: Erlbaum.

Shaw, R., & Turvey, M. T. (1981). Coalitions as models for ecosystems: A realist perspective in perceptual organization. In M. Kubovy & T. Pomerantz (Eds.), *Perceptual organization.* Hillsdale, NJ: Erlbaum.

Sherry, D. F., & Schacter, D. L. (1987). The evolution of multiple memory systems. *Psychological Review, 94,* 439–454.

Shimamura, A. P. (1986). Priming effects in amnesia: Evidence for a dissociable memory function. *Quarterly Journal of Experimental Psychology, 38A,* 619–644.

——— (1989). Disorders of memory: The cognitive science perspective. In F. Boller & J. Grafman (Eds.), *Handbook of neuropsychology.* Amsterdam, Netherlands: Elsevier Press.

——— (1990). Aging and memory disorders: A neuropsychological analysis. In M. L. Howe, M. J. Stones, & C. J. Brainerd (Eds.), *Cognitive and behavioral performance factors in atypical aging.* New York: Springer-Verlag.

Shimamura, A. P., Jernigan, T. L., & Squire, L. R. (1988). Korsakoff's syndrome: Radiological (CT) findings and neuropsychological correlates. *Journal of Neuroscience, 8,* 4400–4410.

Shimamura, A. P., Salmon, D. P., Squire, L. R., & Butters, N. (1987). Memory dysfunction and word priming in dementia and amnesia. *Behavioral Neuroscience, 101,* 347–351.

Simon, H. A. (1962). The architecture of complexity. *Proceedings of the American Philosophical Society, 106,* 467–482.

Singley, M. K., & Anderson, J. R. (1989). *The transfer of cognitive skill.* Cambridge, MA: Harvard University Press.

Slobin, D. (1966). Comments on McNeill's "Developmental Psycholinguistics." In F. Smith & G. A. Miller (Eds.), *The genesis of language.* Cambridge, MA: MIT Press.

Smith, E. E., & Medin, D. L. (1981). *Categories and concepts.* Cambridge, MA: Harvard University Press.

Smith, E. R., & Zárate, M. A. (1992). Exemplar-based model of social judgment. *Psychological Review, 99,* 3–21.

Snodgrass, J. G., & Feenan, K. (1990). Priming effects in picture fragment completion: Support for the perceptual closure hypothesis. *Journal of Experimental Psychology: General, 119,* 276–296.

Squire, L. R. (1986). Mechanisms of memory. *Science, 232,* 1612–1619.

Squire, L. R., & Frambach, M. (1990). Cognitive skill learning in amnesia. *Psychobiology, 18,* 109–117.

Squire, L. R., Amaral, D. G., Zola-Morgan, S., Kritchevsky, M., & Press, G. (1987). New evidence of brain injury in the amnesic patient N. A. based on magnetic resonance imaging. *Society for Neuroscience Abstracts, 13,* 1454–1454.

Stadler, M. A. (1989). On learning complex procedural knowledge. *Journal of Experimental Psychology: Learning, Memory, and Cognition, 15,* 1061–1069.

Stanley, W. B., Mathews, R. C., Buss, R. R., & Kotler-Cope, S. (1989). Insight without awareness: On the interaction of verbalization, instruction and practice in a simulated process control task. *Quarterly Journal of Experimental Psychology, 41A,* 553–578.

Starr, A., & Phillips, L. (1970). Verbal and motor memory in the amnestic syndrome. *Neuropsychologia, 8,* 75–88.

Sternberg, R. J. (1985). *Beyond IQ: A triarchic framework for intelligence.* New York: Cambridge University Press.

Sternberg, R. J. (1986). Inside intelligence. *American Scientist, 74,* 137–143.

Strauss, M. E., Weingartner, H., & Thompson, K. (1985). Remembering words and how often they occurred in memory impaired patients. *Memory and Cognition, 13,* 1507–1510.

Teuber, H. L., Milner, B., & Vaughn, H. G. (1968). Persistent anterograde amnesia after stab wound of the basal brain. *Neuropsychologia, 6,* 267–282.

Thomas, G. J., & Spafford, P. S. (1984). Deficits for representational memory induced by septal and cortical lesions (singly and combined) in rats. *Behavioral Neuroscience, 98,* 394–404.

Thorndike, E. L., & Rock, R. T., Jr. (1934). Learning without awareness of what is being learned or intent to learn it. *Journal of Experimental Psychology, 17,* 1–19.

Tolman, E. C. (1959). Principles of purposive behavior. In S. Koch (Ed.), *Psychology: A study of a science,* Vol. 2. New York: McGraw-Hill.

Tulving, E. (1972). Episodic and semantic memory. In E. Tulving & W. Donaldson (Eds.), *Organization of memory* (pp. 381–403). New York: Academic Press.

——— (1985). How many memory systems are there? *American Psychologist, 40,* 385–398.

Tulving, E., & Schacter, D. L. (1990). Priming and human memory systems. *Science, 247,* 301–306.

Tulving, E., Schacter, D. L., & Stark, H. A. (1982). Priming effects in word-fragment completion are independent of recognition memory. *Journal of Experimental Psychology: Learning, Memory, and Cognition, 8,* 336–342.

Uleman, J. S., & Bargh, J. A. (1989). *Unintended thought.* New York: Guilford.

Vokey, J. R., & Brooks, L. R. (1992). Salience of item knowledge in learning artificial grammars. *Journal of Experimental Psychology: Learning, Memory, and Cognition, 18,* 328–344.

Volpe, B. T., LeDoux, J. E., & Gazzaniga, M. S. (1979). Information processing of visual stimuli in an "extinguished" field. *Nature, 282,* 722–724.

Wagner, R. K., & Sternberg, R. J. (1985). Practical intelligence in real-world pursuits: The role of tacit knowledge. *Journal of Personality and Social Psychology, 49,* 436–458.

——— (1986). Tacit knowledge and intelligence in the everyday world. In R. J. Sternberg and R. K. Wagner (Eds.), *Practical intelligence: Nature and origins of competence in the everyday world* (pp. 51–83). New York: Cambridge University Press.

Warrington, E. K., & Weiskrantz, L. (1968). New method of testing long-term retention with special reference to amnesic patients. *Nature, 217,* 972–974.

——— (1970). Amnesia: Consolidation or retrieval? *Nature, 228,* 628–630.

——— (1974). The effect of prior learning on subsequent retention in amnesic patients. *Neuropsychologia, 12,* 419–428.

——— (1982). Amnesia: A disconnection syndrome? *Neuropsychologia, 20,* 233–248.

Wegner, D. M., & Vallacher, R. R. (1977). *Implicit psychology: An introduction to social cognition.* New York: Oxford University Press.

Weir, R. H. (1962). *Language in the crib.* The Hague: Mouton.

Weiskrantz, L. (1985). Introduction: Categorization, cleverness and consciousness. In L. Weiskrantz (Ed.), *Animal intelligence. Philosophical Transactions of the Royal Society of London, 308 B,* 3–19.

Weiskrantz, L. (1985) (Ed.). *Animal intelligence. Philosophical Transactions of the Royal Society of London, 308 B,* 1–216.

——— (1986). *Blindsight.* New York: Oxford University Press.

Weiskrantz, L., & Warrington, E. K. (1979). Conditioning in amnesic patients. *Neuropsychologia, 17,* 187–194.

Weiskrantz, L., Warrington, E. K., Sanders, M. D., & Marshall, J. (1974). Visual capacity in the hemianopic field following a restricted occipital ablation. *Brain, 97,* 709–728.

Westcott, M. R. (1968). *Toward a contemporary psychology of intuition.* New York: Holt, Rinehart & Winston.

Willingham, D. B., Nissen, M. J., & Bullemer, P. (1989). On the development of procedural knowledge. *Journal of Experimental Psychology: Learning, Memory and Cognition, 15,* 1047–1060.

Wimsatt, W. C. (1986). Developmental constraints, generative entrenchment, and the innate-acquired distinction. In W. Bechtel (Ed.), *Integrating scientific disciplines* (pp. 185–208). Dordrecht: Martinus-Nijhoff.

Wimsatt, W. C. (in press). Generative entrenchment, scientific change, and the analytic-synthetic distinction: A developmental model of scientific evolution. *Philosophy of Science.*

Winter, W., & Reber, A. S. (in press). Implicit learning and natural language acquisition. In N. Ellis (Ed.), *Implicit and explicit language learning.* New York: Academic Press.

Woodruff-Pak, D. S., & Thompson, R. F. (1988). Classical conditioning of the eyeblink response in the delay paradigm in adults aged 18–83 years. *Psychology and Aging, 3,* 219–229.

Woodworth, R. S. (1938). *Experimental psychology.* New York: Holt.

Woodworth, R. S., & Schlosberg, H. (1954). *Experimental Psychology* (Rev. Ed.). New York: Holt, Rinehart & Winston.

Wright, B. M., & Payne, R. B. (1985). Effects of aging on sex differences in psychomotor reminiscence and tracking proficiency. *Journal of Gerontology, 40,* 179–184.

Yoerg, S. I., & Kamil, A. C. (1991). Integrating cognitive ethology with cognitive psychology. In C. A. Risteau (Ed.), *Cognitive ethology: The minds of other animals.* Hillsdale, NJ: Erlbaum.

Young, A. W., & DeHaan, E. H. F. (1988). Boundaries of covert recognition in prosopagnosia. *Cognitive Neuropsychology, 5,* 317–336.

Author index

Subject index